Contents

Introduction

Why use this book?

Exam success depends on two things:

- your knowledge and understanding of the subject matter
- your ability to use that knowledge in the manner that will gain most marks in the examination.

To help you to gain your best possible grade, this book aims to:

- set out the subject content essential to the CCEA GCSE Geography course
- give you hints and revision tips that will help you to understand and memorise the material
- advise you on the best way to approach various types of exam questions.

Each chapter covers one of the six themes making up the GCSE course, as well as providing tasks, knowledge tests and exam questions from past papers so that you can check, as you go along, how much you understand and can remember. The answers are set out in the final chapter. The glossary, pages 116–20, provides clear and concise definitions of the key ideas required for each topic. This section is important as you will almost certainly be asked to define some of them in the examinations.

Structure of the examination

In addition to the controlled assessment (worth 25% of the overall marks), which is completed and assessed in school, there are two examination papers, each worth 37.5% of the total marks.

Paper 1	Paper 2
Three physical geography themes: • The Dynamic Landscape • Our Changing Weather and Climate • The Restless Earth	Three human geography themes: • People and Where They Live • Contrasts in World Development • Managing Our Resources
You must answer all three questions; one on each theme	You must answer all three questions; one on each theme
Each question is made up of several parts, including both short answers and extended writing	Each question is made up of several parts, including both short answers and extended writing
Time: 1.5 hours Spend 45 minutes on question 1 and 20 minutes on each of the other two questions	Time: 1.5 hours Spend 45 minutes on question 1 and 20 minutes on each of the other two questions

Further details of the specification, including the requirements for each of the six themes, can be downloaded from the CCEA website at www.ccea.org.uk/geography.

Revision techniques

Here are some useful tips about revision:

- Ideally, revision should be ongoing throughout the course. Don't leave it all to the days just before the exam.
- Case studies, in particular, should be memorised as you go along so that facts about each one are clear in your mind before you study the next.
- Revision is *not* just re-reading your notes or the textbook.
- Revision should involve reworking the subject matter, perhaps into a spider diagram or by summarising into brief bullet points.
- Next, you have to memorise the material by repeating it to yourself, explaining it to someone else, writing a list or making a poster.
- Test yourself, to see how much your memory has retained, by writing out the bullet points or list, re-drawing the diagram or explaining it all to someone – this time without the help of your notes or book.
- Visual forms of revision, such as spider diagrams and learning maps, can be a big help. They let you picture the key points, arrange them under headings and see connections.
- Case studies need careful revision, with facts, figures and places committed to memory. Writing the title of each one with bullet points covering the main ideas (including the facts, figures and place names) on a card will help you to concentrate on this aspect of the course. Gradually, you will build up a collection of cards, ready for last-minute revision on the eve of the exam.

Do →	**Rework** →	**Memorise** →	**Write an answer**
Stage 1	*Stage 2*	*Stage 3*	*Stage 4*
• Do classwork. • Do homework. • Do make notes highlighting key words and relevant facts: places, figures and names.	• Change notes into spider diagrams or learning maps. • Add own ideas/ theory and make links with previous notes. • Discuss your work with the teacher or a friend. • Make a summary box and structure the information under meaningful headings. • Find new ways of thinking about something, such as using thinking skills.	• Commit to memory by repetition. • Say the information out loud or make up a rhyme or tune. • Explain your topic to a friend. • Put the information up on posters around your room and move around when learning. • Take small breaks. • Close your notes. • Recall information by writing it out.	• Select the information which is relevant to answer a GCSE question. • Compose an answer to the question.

Examination techniques

Command words

To answer exam questions correctly, it is important to be sure what the examiner is really asking. Read the question carefully and underline the command words – these are words such as *state*, *describe* or *explain*. They tell you what to do in your answer. If you explain when asked to describe you will earn no marks, even if what you write is otherwise correct.

The following table gives the meanings of some of the command words you will meet.

Command word	Meaning
State	A short answer, presenting a fact or facts (for example, the temperature in January, taken from a graph)
Describe	A descriptive answer *without* trying to explain When describing a *graph*, it is important to *quote figures* When describing a *map*, it is important to mention *place names*
Explain	Give a reason or reasons
Describe and explain	Make descriptive statements and give the reasons why (for example, describe the pattern of rainfall shown on a map and explain why it falls there)
Label	Add labels to a diagram
Complete	Add information to a graph or a table so that it is complete
Match	Match statements that have been presented in the form of 'heads and tails'
State the meaning	Usually used for definitions. You need to show that you know what the term means
Suggest	This is used when there may be more than one correct answer and any relevant one is acceptable

Different types of questions

Recall questions

These are designed to test your knowledge. Definitions of key ideas are in this category, for example, 'State the meaning of the term earthquake'.

Data response questions

These questions may be based on a table of data, graph, Ordnance Survey (OS) map, photograph, sketch map, weather map, cartoon or newspaper article. They test what you can observe from the map or diagram and what you understand from what you have seen. These questions can make it easier to gain marks than with recall questions, since they provide you with visual clues, so try to make the best use of them.

- *Tables and graphs*. If you are asked to describe a graph or table, you need to put into words what it shows, remembering to quote some figures to support your statements.

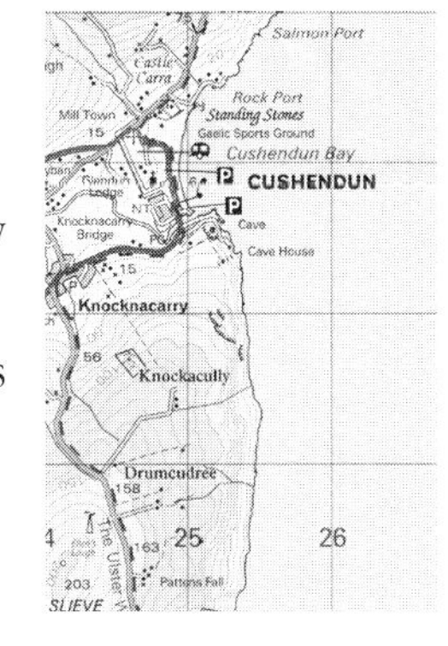

- *OS maps*. An OS map is included with every examination, either with paper 1 or 2. A key to the symbols is printed alongside the map and the scale will be 1:50,000. This means that 2 cm on the map represents 1 km on the ground, and every grid square is 1 km × 1 km. You need to be able to:

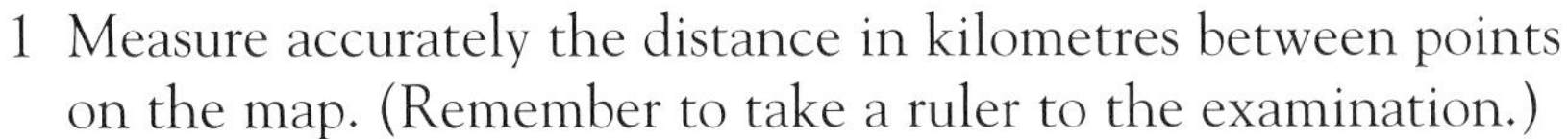

  1 Measure accurately the distance in kilometres between points on the map. (Remember to take a ruler to the examination.)
  2 Use compass directions to say, for example, that farm A is north-east of village B. Use only the eight compass points – N, NE, E, SE, S, SW, W, NW.
  3 Read the height above sea level from contour lines, spot heights or triangulation pillars and interpret relief shown by contour patterns.
  4 Find features on the map using four-figure and six-figure grid references. Remember that a four-figure reference describes a whole *square* on the map. The first two figures indicate the vertical line and the final two indicate the horizontal line that makes the bottom left-hand corner of the square in question. A six-figure grid reference describes a *point* on the map, with the third and sixth figures telling you how many tenths of the way across or up the square you need to go.
  5 Apply your knowledge and understanding of the various themes (settlement, economic activity, coasts and rivers and so on) to evidence presented in the OS map.

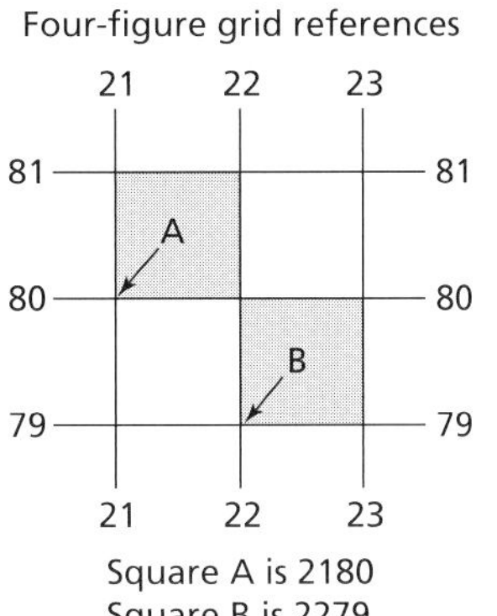

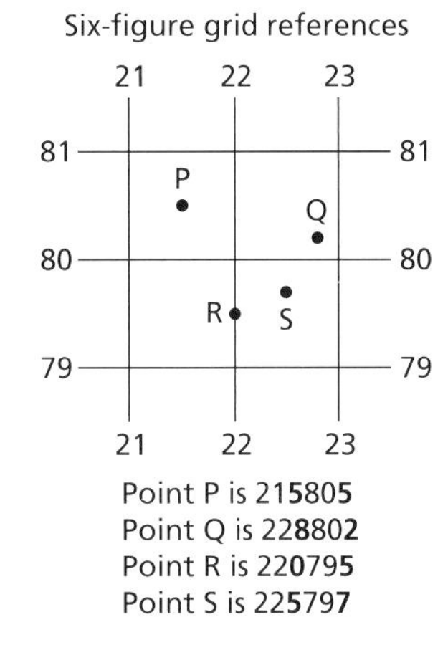

- *Photographs*. In an exam you may be asked to match up a photograph with a location on an OS map or to comment on the geographical feature shown in the picture. Try to make full use of the evidence presented in the photograph.
- *Weather maps*. Look at the pattern of the isobars. If the highest pressure is in the centre, it shows an anticyclone with descending air, calm conditions and clear skies. If the lowest value is in the centre it shows a depression with rising air, and cloud and rain along the fronts. Take note of the time of day and date stated next to the map as these help to explain the situation presented. By looking at the wind direction you may be able to suggest what air mass is influencing the area. A key is provided for the symbols used on the map, for example cloud cover and wind speed.

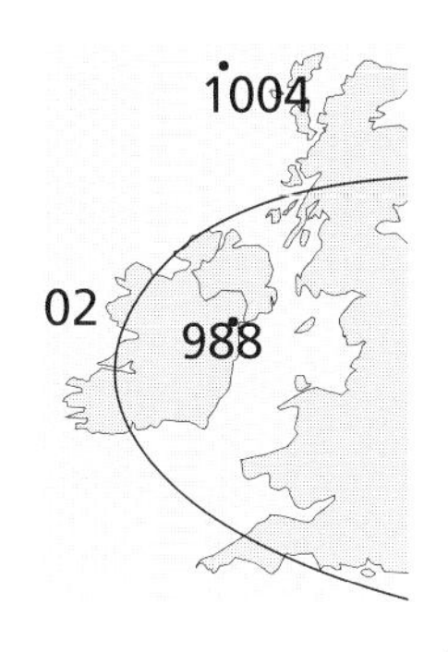

Levels of response questions

These are questions worth 6 marks or more, and they require extended written answers. For a 6-mark question, the mark scheme may be:

- Level 1 1–2 marks a correct, but simple answer
- Level 2 3–4 marks a fuller answer, developing the basic idea
- Level 3 5–6 marks an answer showing greater depth of understanding.

Examples of 'levels of response' questions include 'Explain one impact …' or 'Explain why …' . There are likely to be 3 marks for each full explanation. To earn these 3 marks you need to:

- make a statement
- give a reason or consequence of the statement
- elaborate on the reason or consequence. This can be further detail or an actual named example.

Here is how you might answer the question: 'Study the diagram showing the original site of Belfast. Explain why this site would have been attractive to settlers. You should give **two** reasons in your answer.' [6]

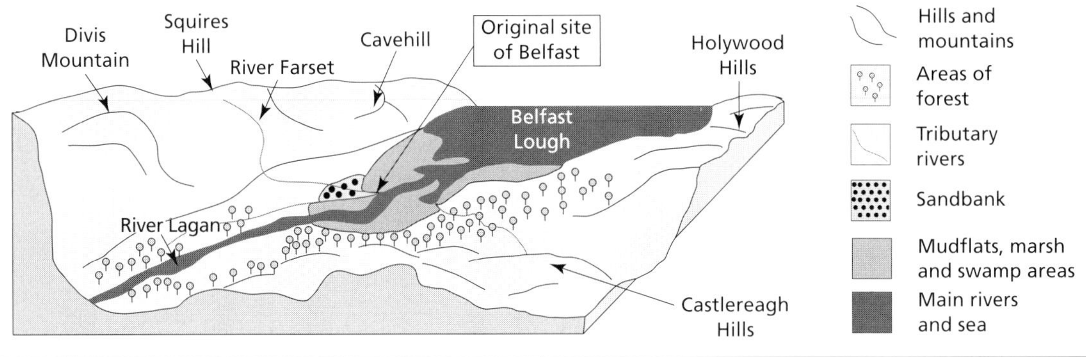

First reason	Second reason
• Statement: People can get wood from the woodland ✓ (1) • Consequence: … which they need to cook their food ✓ (1) • Elaboration: … or to build their houses ✓ (1)	Either (for 2 marks): There is flat land ✓ (1) which could be used for farming ✓ (1) Or (for 3 marks): The flat land between Divis Mountain and the Castlereagh Hills was suitable for farming to provide food for the settlers

Case studies

Case study questions are levels of response questions in which you need to show that you are writing about a real place. The facts should be as specific as possible, for example, mentioning countries by name rather than referring to LEDCs or 'Africa'. For top marks you should aim to include at least two facts (figures, names or places) that are specific to the place you are writing about. For example, when writing about the Kobe earthquake case study, '5500 people died' is worth more marks than 'lots of people killed' and 'unemployment when Mitsubishi factory closed' is better than 'jobs lost because factories closed'.

A Understanding Our Natural World

The Dynamic Landscape

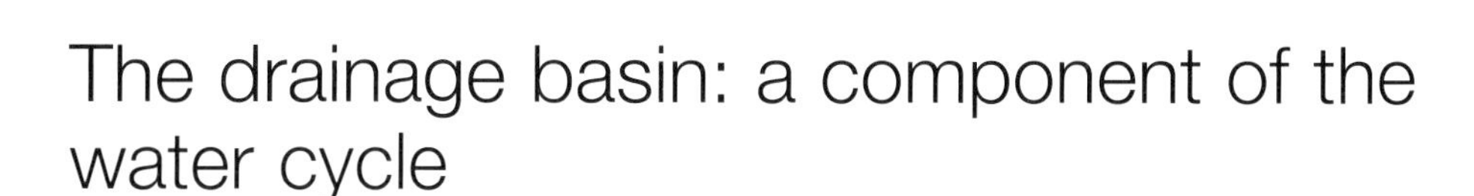

The drainage basin: a component of the water cycle

You need to be able to:

- understand how water moves around in the drainage basin
- identify and define parts of a drainage basin.

The **water cycle** is the way water is evaporated from the sea, goes through the air and flows back to the sea through rivers or the ground.

A **drainage basin** is the area of land drained by a river and its tributaries. In other words, any rain that falls in a particular area of land will end up in one particular river.

The rain may have a very eventful journey before it reaches the river. You need to understand the different parts of that journey. We sometimes talk about the parts of the journey as a system – just like the way food goes through our digestive system.

All systems have:

- things which go into them (inputs)
- ways of moving something from one place to another (transfers)
- places where things are stored (stores)
- things which come out at the end (outputs).

	Name	Meaning
Input	**Precipitation**	Any water falling from the sky: rain, snow, sleet, hail
Stores	**Interception** by vegetation	Leaves and grass catch raindrops as they fall, and store them. Try sheltering under a tree next time you get caught in the rain! But don't stay too long – if there's too much water stored on a leaf it can fall to the ground.
Transfers	**Surface run-off/overland flow**	Water running over the surface of the ground.
	Infiltration	Water sinking into the soil
	Through-flow	Water flowing through the soil
	Percolation	Water sinking down through the rock
	Groundwater flow	Water flowing slowly from the rock into the river
Outputs	River **discharge**	Water flowing away in the river
	Evaporation	Water turning into water vapour in the air

Features of a drainage basin

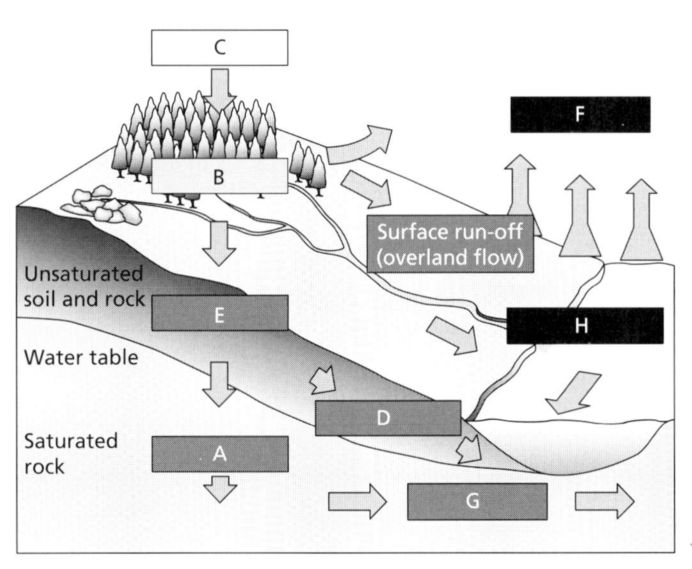

◀ **Figure 1** The drainage basin system

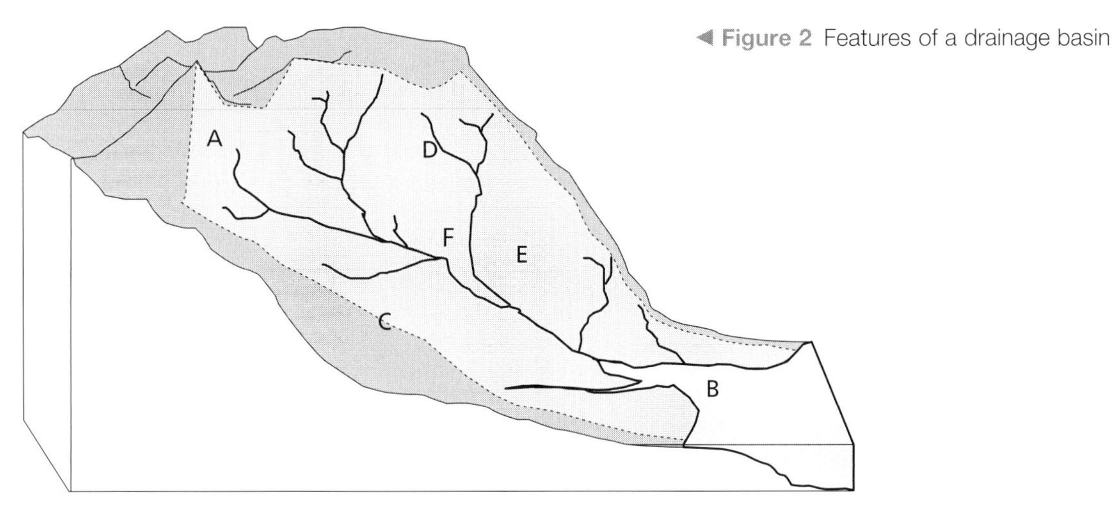

◀ **Figure 2** Features of a drainage basin

Get Active

1 Match the words from the middle column of the table on page 1 to the boxes A–H on Figure 1.

2 Match the letters A–F on Figure 2 to the following labels:

Drainage basin – the area of land drained by a river and its tributaries.
Source – the place where a river starts.
Tributary – a stream flowing into a river.
Mouth – where a river flows into the sea.
Watershed – the highest point all round a drainage basin.
Confluence – where two streams or rivers meet.

River processes and features

You need to be able to:

- explain what changes occur along the long profile of a river, and why
- understand erosion, transportation and deposition
- understand how waterfalls, meanders and floodplains are formed, using diagrams
- use aerial photographs and OS maps to identify river features and land uses.

Changes along the long profile of a river

The long profile of a river means its shape from the source to the mouth. Imagine cutting down through the land to be able to see the whole river from source to mouth.

▲ **Figure 3a** Profile of a face

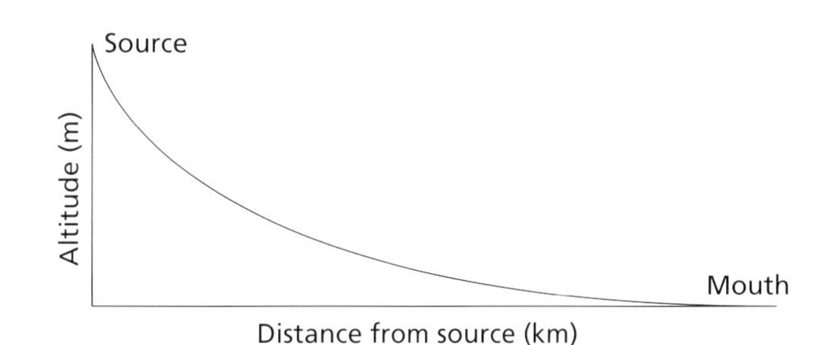

▲ **Figure 3b** Sketch of long profile of river

	Gradient	Depth	Width	Discharge	Load
Meaning	The steepness of the slope the river flows down	Measure from top of water to river bed. Take average across river	Distance from one side of the river to the other	Amount of water passing a point in a certain time – cumecs (cubic metres of water per second)	The material a river is carrying – mud, sand, pebbles, rocks
Change as you go downstream	Gets less steep	Gets deeper	Gets wider	Increases	Particles get smaller and more rounded
Why?	The river does more downwards erosion near the source, and more sideways erosion near the mouth	The river erodes downwards as it travels (vertical erosion)	The river erodes sideways as it travels (lateral erosion)	More water flows into the river from each tributary Water flows faster with less friction	Particles knock against each other and break each other up. Sharp angular edges get knocked off

▲ **Figure 4** River channel changes along the long profile

Get Active

Complete the table below using the following words.

angular deep gentle high large low narrow rounded shallow small steep wide

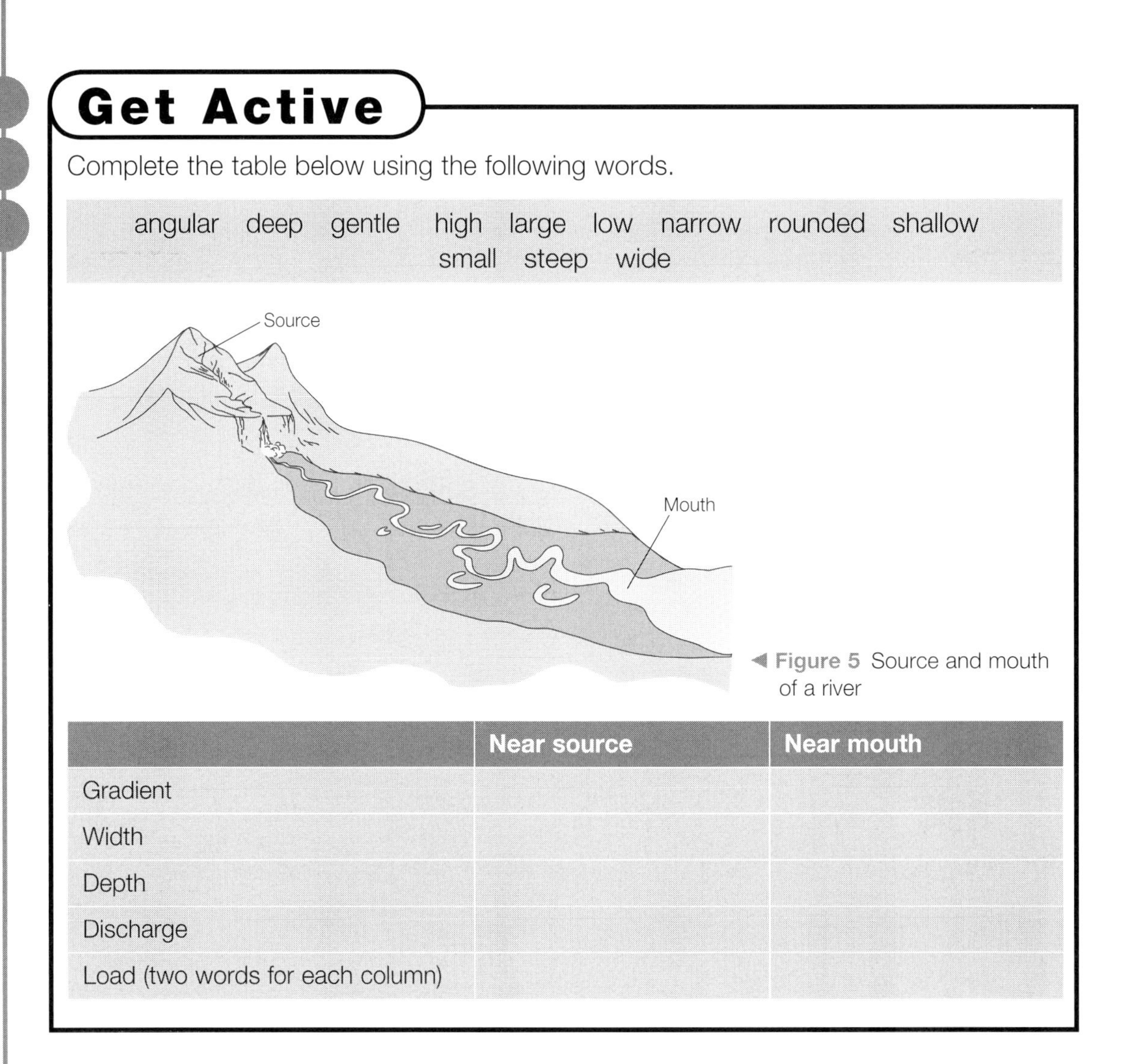

◀ **Figure 5** Source and mouth of a river

	Near source	Near mouth
Gradient		
Width		
Depth		
Discharge		
Load (two words for each column)		

● Processes of erosion, transportation and deposition

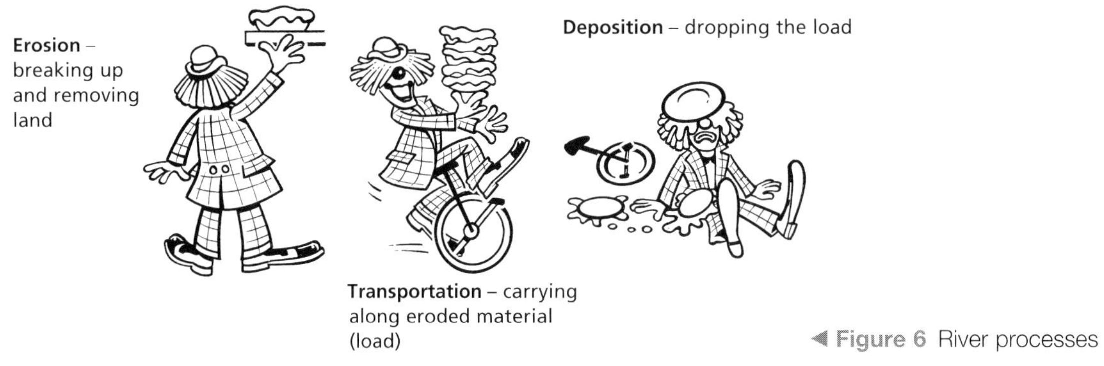

◀ **Figure 6** River processes

A more detailed explanation of the processes of erosion can be seen in Figure 7 – the same processes work at the coast and in rivers.

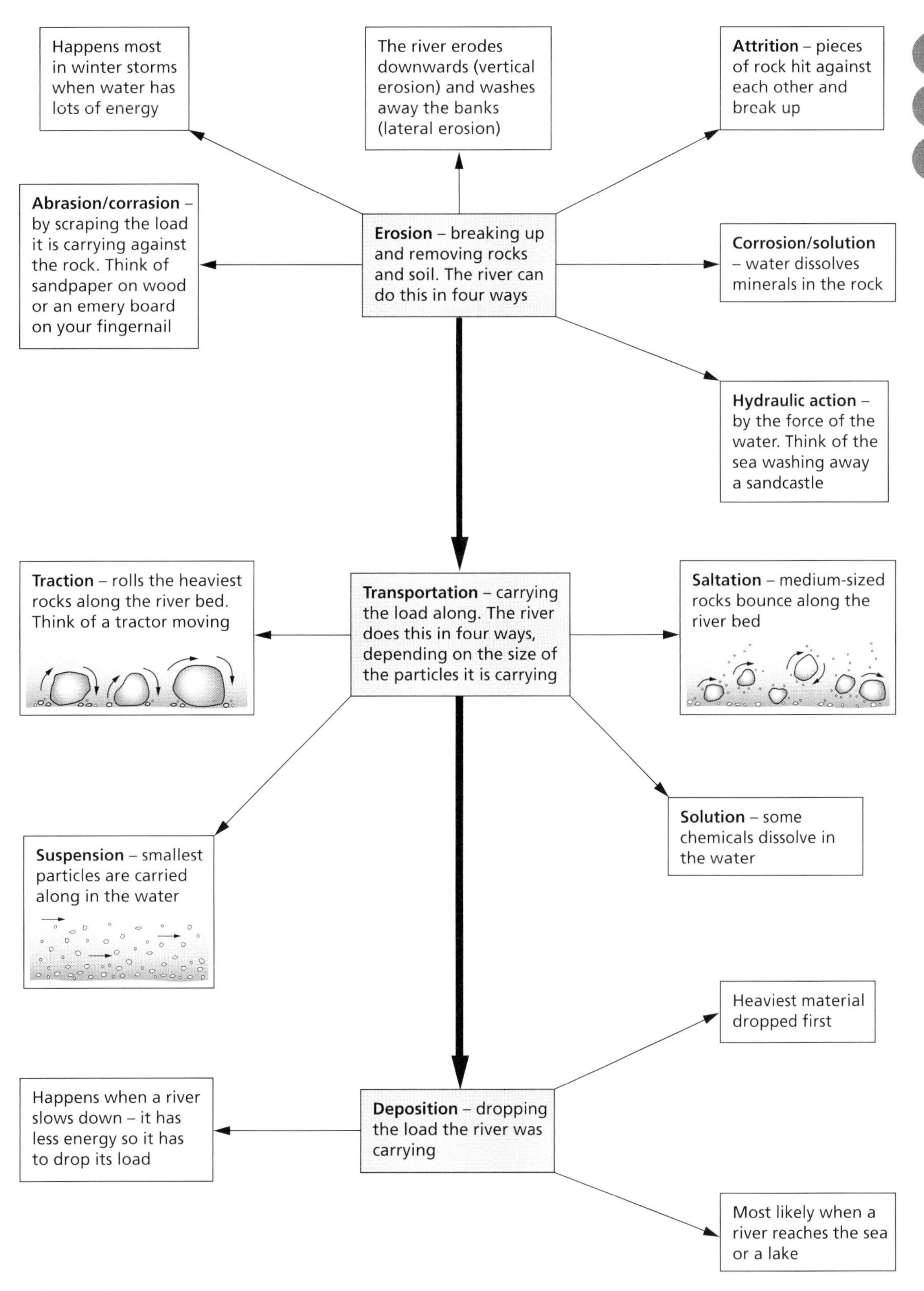

▲ **Figure 7** Processes carried out by a river

Get Active

1 For each of the following descriptions, decide what process is happening:
 a Large pieces of rock are broken off the edge of the river bank.
 b The river looks very brown because it is carrying lots of mud.
 c During a flood the river gets wider.
 d There is a tiny beach at the edge of the river where it flows into a lake.
 e If you paddle barefoot in the river you may get stones hitting against your feet.
 f Pebbles on the river bed are sharp near the source and smooth near the mouth.

2 Go to Geography At The Movies website: www.gatm.org.uk. Follow the links to Archive Pages then Rivers, and find 'Sticky does River Processes'. This is a brilliant clip using cartoons to explain all the processes you need to know. Watch the clip, then try to draw your own cartoons to illustrate all the words in bold in Figure 7. The funnier you make them, the more likely you are to remember what they mean!

Formation of river landforms

You need to be able to use annotated diagrams to explain this, and identify the landforms on maps and aerial photographs.

Formation of a waterfall

- A waterfall is formed when there is a layer of **hard rock** on top of a layer of **soft rock**.
- The river erodes the soft rock more easily, so there is a step in the river bed. Eventually this becomes deeper, making a waterfall.
- **Hydraulic action and abrasion make a plunge pool** at the bottom of the waterfall.
- More erosion **undercuts** (or cuts under) the hard rock, leaving it hanging over the plunge pool.
- **The overhanging hard rock falls into the plunge pool**, and the position of the **waterfall moves backwards**.

On an OS map most waterfalls are marked with a written label saying 'Falls', or 'Waterfall'. Have a look in the main textbook at page 6, grid square 1931 and 2024.

Get Active

1 Match the words in bold (above) with the numbers in Figure 8 to label the diagram.

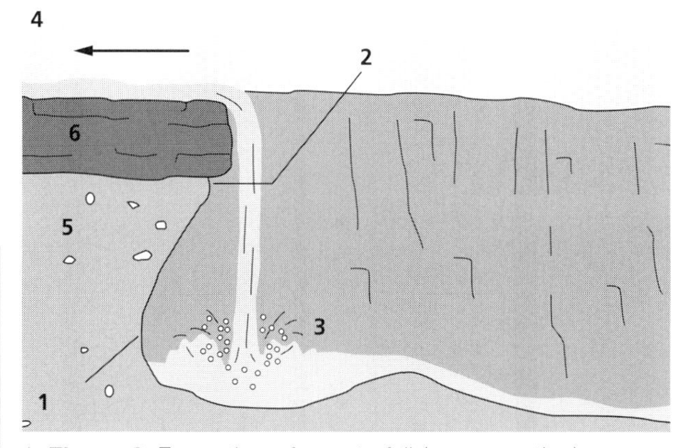

▲ Figure 8 Formation of a waterfall (cross-section)

Formation of a meander

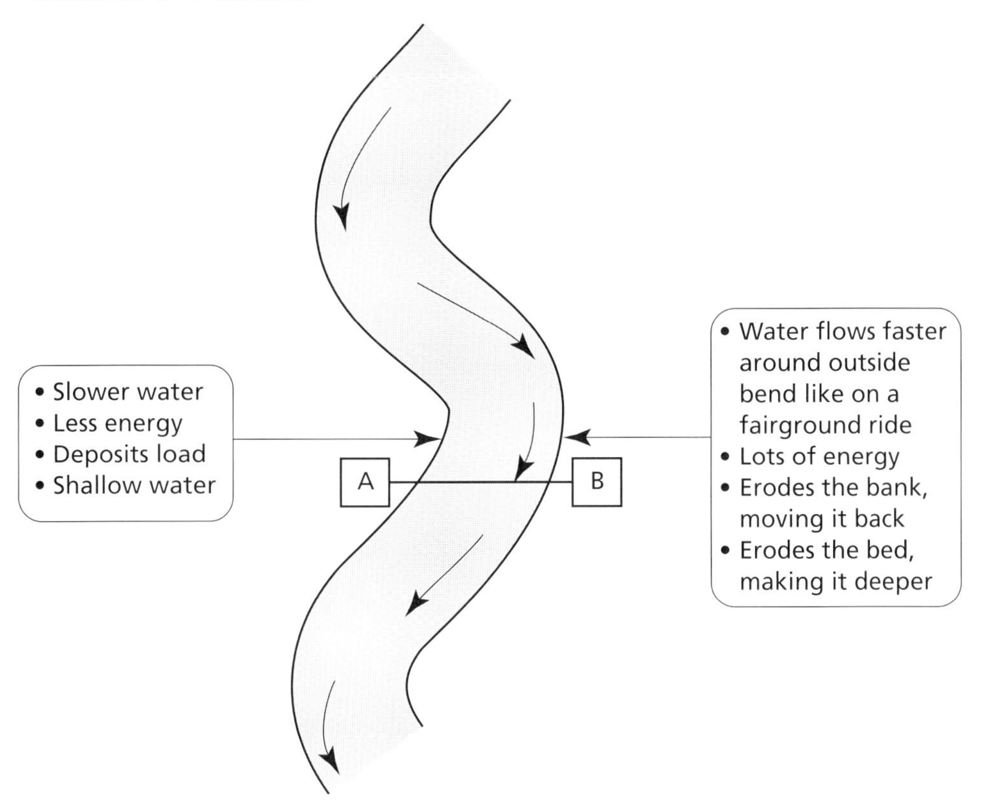

▲ **Figure 9** Formation of a meander

Figure 10 shows what it would look like if you could cut through the river along the line marked A–B (a cross-section). In an exam you might have to label any of these features on a diagram or a photo.

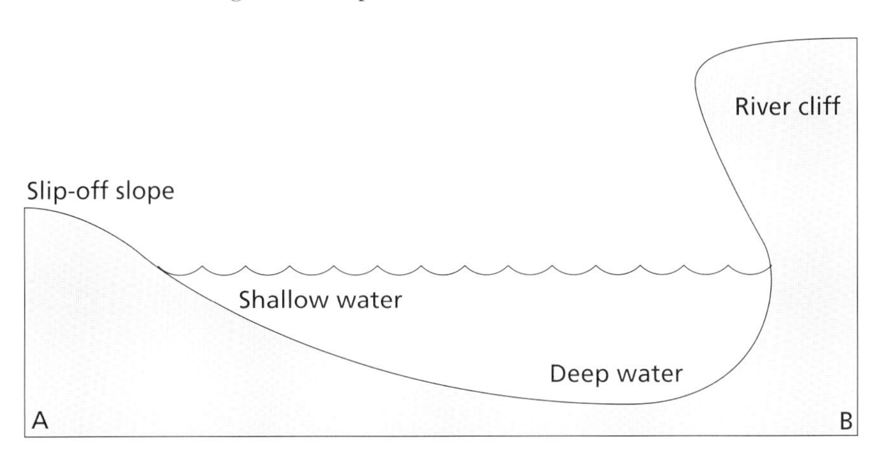

▲ **Figure 10** Cross-section through channel along the line A–B

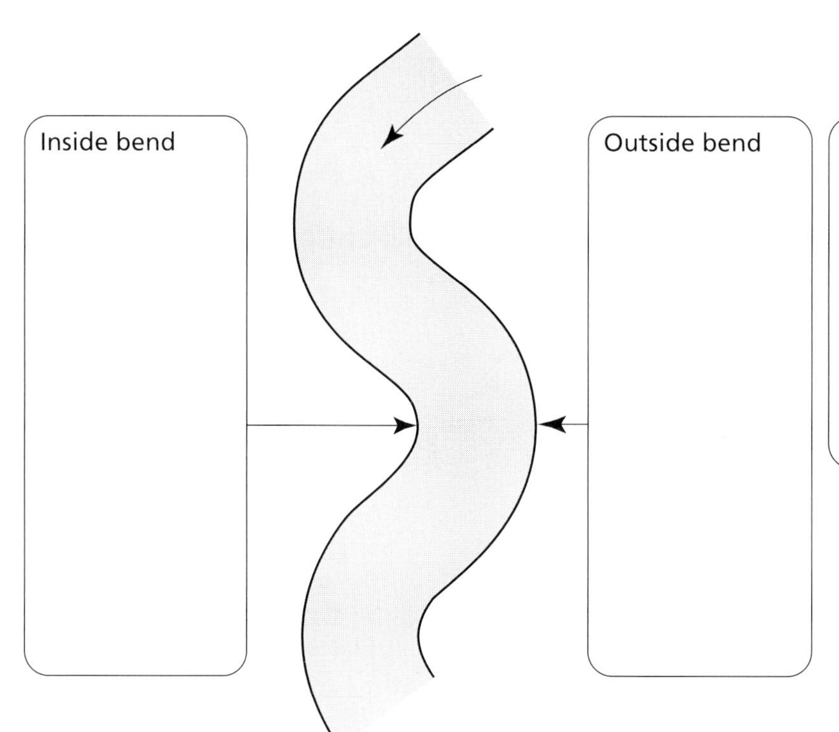

▲ Figure 11

Get Active

1 Check you have understood the main facts about meanders by copying and completing Figure 11.
   - Choose appropriate words from the word box to label what happens at each side of the river bend.
   - Draw arrows to show where the water flows fastest in the river.

2 Draw a cross-section diagram of C–D from Figure 12. Label:
   - Fast-flowing water.
   - Slow-flowing water.
   - Deep water.
   - Shallow water.
   - Erosion.
   - Deposition.
   - River cliff.
   - Slip-off slope.

3 Look at page 7 in *Geography for CCEA GCSE, Second Edition*. In the north of the map (Figure 4), west of Cushendun, you will find a river with several meanders. Draw the shape of the river, and label the places where **a** erosion and **b** deposition will take place.

4 Look at page 12 in *Geography for CCEA GCSE, Second Edition*. Figure 11 shows an aerial photograph of the Mississippi River. You should be able to spot areas where deposition has happened. What colour do you see there?

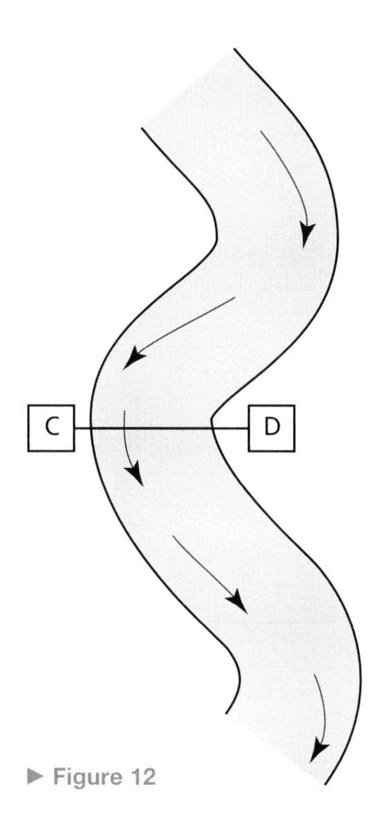

► Figure 12

Formation of a floodplain

Flat land either side of a river will be covered in water if the river bursts its banks: this is called a floodplain.

Where there are meanders, the river flattens the land by eroding it and depositing sediment on it, making a flat floodplain. If the river gets too full and bursts its banks, water floods over the floodplain.

The water slows down, loses energy, and deposits load (called sediment or alluvium) which is fertile (good for growing crops).

In an exam you might have to label (annotate) a diagram or photo.

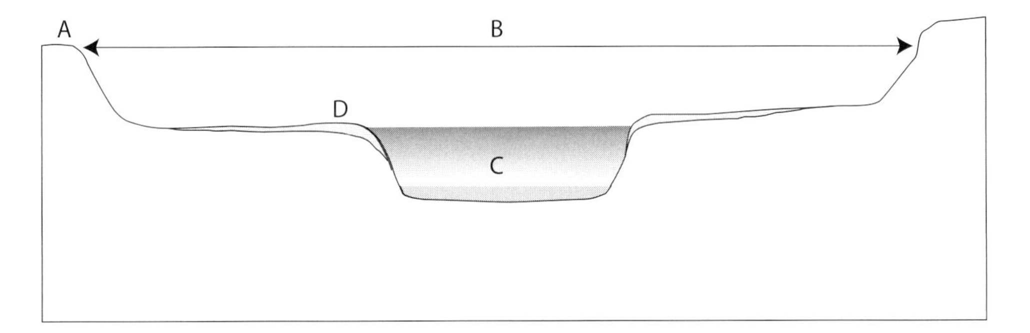

▲ **Figure 13** Cross-section of a floodplain

Get Active

1. Use the definitions below to help you label letters A–D in Figure 13.
   River channel – where the river flows.
   Bluff – remains of higher land, where the rest has been eroded.
   Floodplain – flat area covered in water in a flood.
   Deposition – process where flood waters drop the material they are carrying.
2. Look at page 7 of *Geography for CCEA GCSE, Second Edition*. The floodplain of the Glenariff River stretches to the south-west of Glenariff or Waterfoot, grid square 2425.
   - **a** What colour is the land in the floodplain shaded on the map?
   - **b** What height is the land? (look at the contour lines)
   - **c** How can you tell it is flat?
   - **d** What do you think are the land uses on this floodplain.
   - **e** What problems will there be if the river floods on its floodplain?

2006 Past Paper Exam Questions (Higher Tier)

2 (a) Study Table 1 which shows how the Whitewater River in the Mourne Mountains changes downstream. Answer the questions which follow.

Table 1

Distance from source (km)	Width of river channel (m)	Depth of river channel (m)	Size of load. Longest axis (cm)
1	1.27	0.05	14
17	12.2	0.24	7

(i) Describe how the river channel changes downstream. [4]

(ii) The load is smallest near the mouth of the river. State fully **one** reason why this is so. [3]

This is a data response question – you can get marks very easily if you remember to use the data properly.

You have to identify changes in the river channel (**not** the load) and for full marks you need to quote figures. For example, 'The river gets wider from 1.27 m to 12.2 m. It also gets deeper from 0.05 m to 0.24 m'.

State fully one reason why this is so – make sure you give statement (S), consequence (C) and elaboration (E).

For example, 'Rocks hit against each other and the river banks and river bed as they go downstream (S). This makes them become smaller as bits are knocked off them (C). This is called erosion (E).'

Coastal processes and features

You need to be able to:
- understand the difference between destructive waves and constructive waves
- understand the impact that each has in changing the coastline.

Waves

Coastlines are dynamic – this means that they change through time. Cliffs can crumble and collapse; beaches may have more or less sand this year than last. Wave action is the main process that causes these changes. Destructive waves attack the coast whereas constructive waves add sediment to the shore.

Waves are the shapes on the surface of the sea, caused by wind blowing over it. You can create similar waves if you spread a cloth over a table and push the cloth with your hand: a gentle push produces gentle ripples while a strong push creates higher waves.

The type of wave that breaks on the shore can be either **constructive** or **destructive**, depending on the wind speed and how far the waves have travelled. A strong gale force wind blowing for 5000 km across the Atlantic Ocean causes much higher waves than a breeze blowing 150 km across the Irish Sea. When the wave breaks, water surges up the beach (called swash), and then returns down the beach to the sea (called backwash). Both of these movements are important.

	Destructive waves	Constructive waves
Size	High and close together	Low and far apart
Frequency	Frequent, up to 15 per minute	Less frequent, 6–9 per minute
Season	Common in winter (storm waves)	Common in summer
Effects	Stronger backwash than swash Drags sand and pebbles out to sea Erodes the coast	Strong swash and weak backwash Pushes sand and pebbles up the beach Causes deposition which builds up the coast

Get Active

Figures 14 and 15 show destructive and constructive waves. Match the labels below with the letters A–G on the diagrams.

- Higher wave
- Waves far apart
- Water sinks into beach reducing backwash
- Strong backwash pulls sediment down the beach
- Lower wave
- Waves close together
- Strong swash pushes sediment up beach

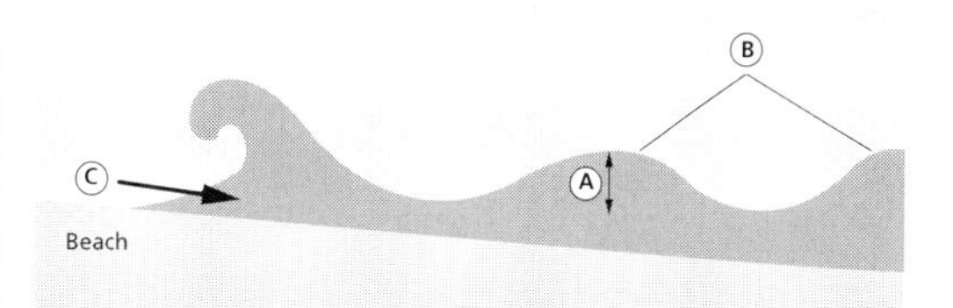

▲ **Figure 14** Destructive waves

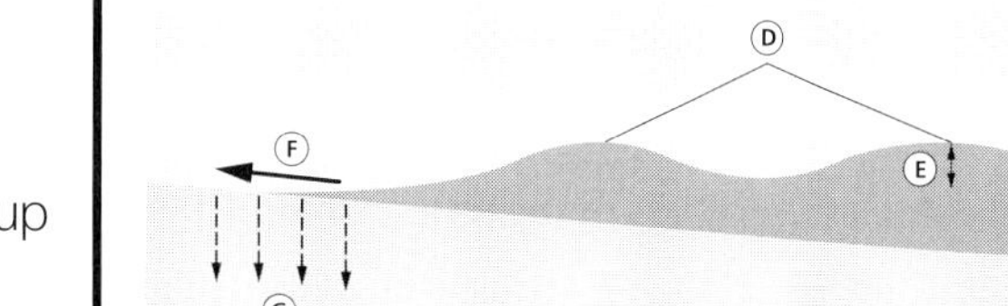

▲ **Figure 15** Constructive waves

Coastal erosion, transportation and deposition

You need to know and understand processes of:
- erosion: abrasion (corrasion); attrition; solution (corrosion); hydraulic pressure
- transportation: longshore drift
- deposition.

Erosion processes

These four processes are similar to the ones you have learned for rivers, but can have alternative names.

1. **Hydraulic pressure** is the force of the waves alone, especially in storm conditions.
2. **Abrasion (corrasion)** is the sand-papering action of water carrying sand and pebbles – it smoothes and wears away the rocks where waves hit at the base of a cliff.
3. **Solution (corrosion)** is the chemical action of seawater dissolving rocks such as limestone.
4. **Attrition** occurs as pebbles transported by waves hit against each other, breaking into smaller and rounder particles.

Get Active

1. Identify the three erosion processes shown in Figure 16a–c.
2. Draw quick sketches of the three diagrams and select appropriate labels for 1–4 from the list below.
   a Air compressed (squashed) by advancing wave.
   b Wave carries sand and pebbles which wear away the rock.
   c Compressed air expands after wave breaks and loosens blocks of rock which fall into the sea.
   d Pebbles crash against each other becoming smaller and rounder.

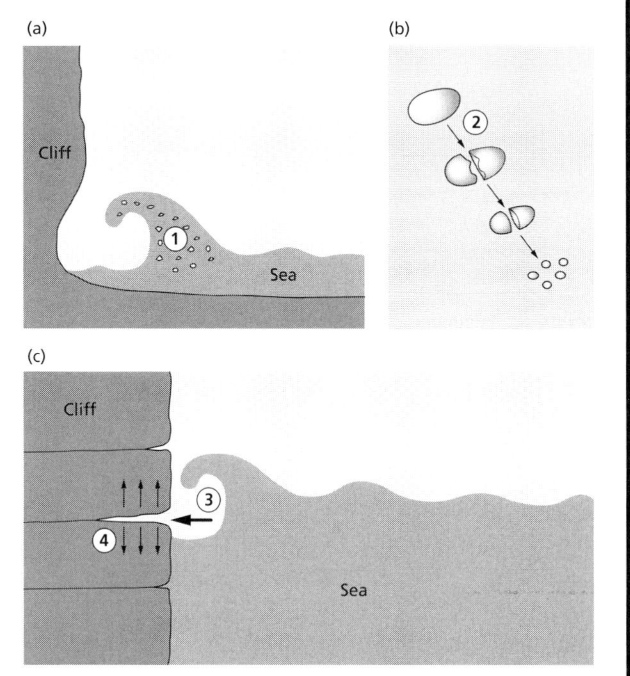

▲ **Figure 16** Coastal erosion processes

Transportation

The pieces of rock eroded by the sea are transported away by wave action. When waves break normally on a beach, each swash carries this material up the beach and the backwash sweeps it back out to sea, as shown in Figure 17a.

If, however, waves approach the shore at an angle, the swash carries material up the beach at the same angle, but the backwash returns it straight back to the water's edge because of gravity (see Figure 17b). This zig-zag movement of sand or pebbles along the coast is called **longshore drift**.

Deposition

When waves are constructive they deposit the load they carry onto the shore. This often happens in a sheltered bay, between two headlands.

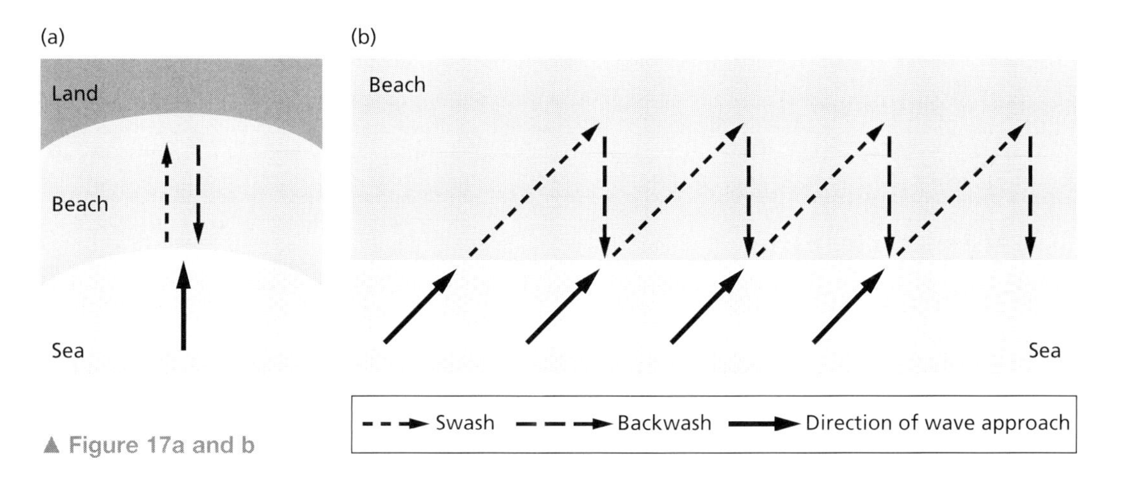

▲ Figure 17a and b

● Landforms resulting from erosion and deposition

You need to be able to:

- explain the formation of erosional landforms: cliff, wave cut platform, cave, arch and stack
- explain the formation of depositional landforms: beach and spit.

You should be able to refer to places when you write about this.

Erosion landforms

When destructive waves approach a rocky headland, the processes of **hydraulic pressure** and **abrasion** erode a wave cut notch. Through time, this notch gets bigger and the unsupported rock above it is undercut and collapses into the sea, forming a **cliff**. Continued erosion and repeated collapse causes the cliff line to retreat. This leaves an almost level area of rocks and rockpools known as a **wave cut platform** at the base of the cliff.

You can see a wave cut platform at Portstewart as you walk along the cliff path towards the Strand.

Erosion attacks the rock where it is weak, to form **caves**. Hydraulic pressure is very effective in making caves bigger as air gets trapped in the cave by advancing waves and is forced into cracks in the roof and at the back of the cave. Over time this means the cave can extend right through the headland to form an **arch**. When the roof of the arch collapses it leaves an isolated **stack** which can be eroded to leave only a stump.

Get Active

1 Make a quick sketch of the diagrams in Figure 18 and select appropriate labels for 1–4 from the list: (a) cliff, (b), wave cut platform (c) cliff moves backwards, (d) wave cut notch.

2 Copy Figure 19 and choose the correct labels for 1–4 from the list: (a) cave, (b) stack, (c) crack or weakness in rock, (d) arch.

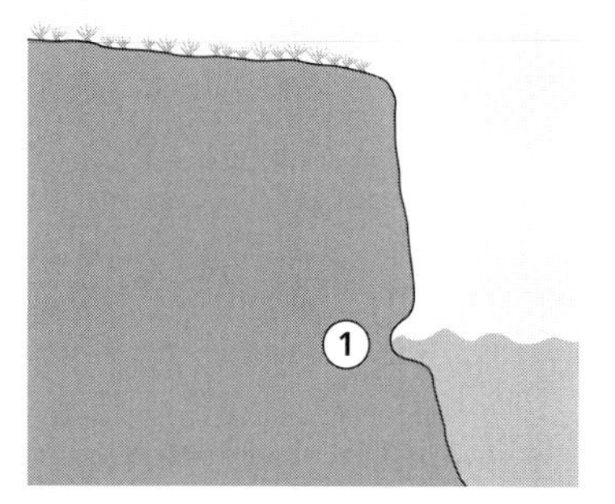

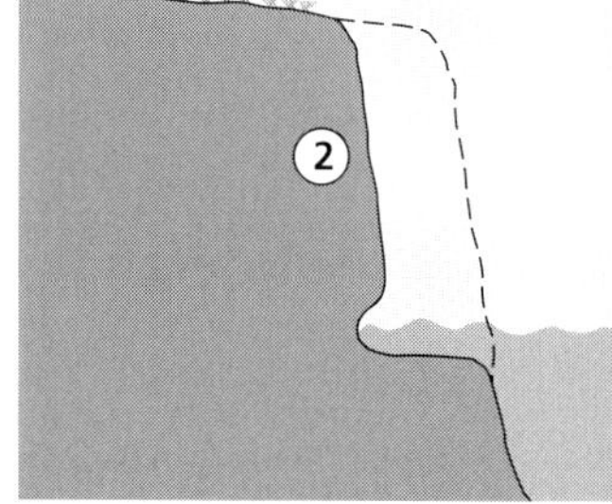

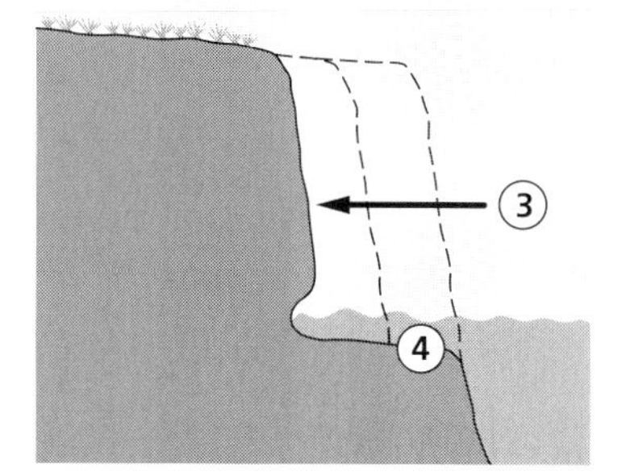

▲ **Figure 18** Cliff retreat

▲ **Figure 19** Caves, arches and stacks

Deposition landforms

A **beach** forms where sand or pebbles build up in a sheltered position on a coastline. It is built by constructive waves which, over time, add sediment with their swash – the backwash is too weak to remove all of it. On most beaches the swash and backwash both operate straight up and down the beach so sand or pebbles do *not* move along the shore (see Figure 17a). Pebble beaches are usually steeper than sandy beaches.

A **spit** is an extended beach that forms a 'finger' of land sticking out into the sea. It is formed when sediment is carried along by **longshore drift** and deposited out into the sea where the coastline turns a corner. Spurn Head in England is a classic example of a spit.

Get Active

Study Figure 20. Select appropriate labels from the list below to complete the explanation.

a Deposition in sheltered area behind spit.
b Sediment moves along the beach by longshore drift.
c Spit grows as sediment is added.
d Waves approach coast at an angle.

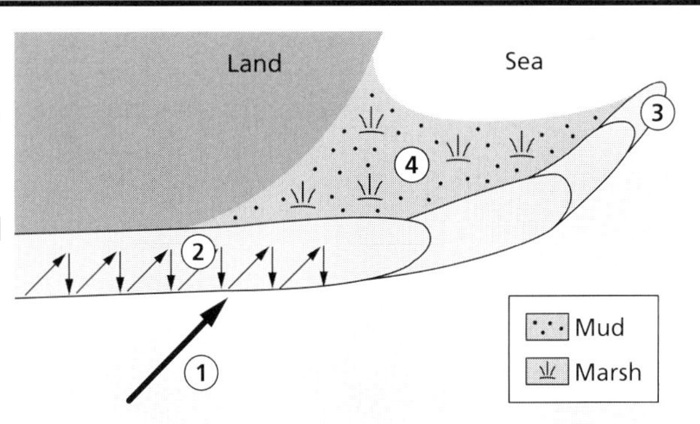

▲ **Figure 20** The formation of a spit

Coastal features and land uses on aerial photographs and OS maps

You need to be able to:

- identify coastal features and landforms from OS maps and aerial photographs.

Get Active

1 Using the OS map on pages 6–7 of your textbook, *Geography for CCEA GCSE Second Edition*, match the grid references **a–d** with the correct coastal features (1–4).

Grid references	Coastal features
a 242257	**1** Cave
b 246286	**2** Arch
c 2426	**3** Beach
d 2532	**4** Cliffs

2 Study the Google image of Spurn Head on page 22 of your textbook. In what direction must sediment be moving by longshore drift to form and extend this spit?

Sustainable management of rivers

If people are going to live near rivers they need to make sure that what they do to that river is sustainable. For example, if building a dam causes problems further down the river, then it is not a sustainable form of management.

● Physical and human causes of flooding

You need to:
- explain the causes of a specific flood event in the British Isles, using detailed facts and figures.

Flooding occurs when the water in a river channel is higher than the river bank, so it overflows. Most rivers flood regularly and this may result from physical causes (such as heavy rainfall) or from human activity (such as building on the floodplain).

Case study: Causes of flooding – River Derwent, March 1999

Physical causes:	**Human causes:**
• Heavy rainfall over the North York Moors, 250 mm in 12 days, as much as usually falls in 3 or 4 months. • Ground in the Derwent drainage basin was already saturated (completely soaked) by previous rain so could hold no more water.	• The peat of the North York Moors used to store water like a sponge for days before releasing it to the rivers, but people had cut and removed lots of peat. • Houses have been built on the floodplain, at Malton for example, making the ground there impermeable (water cannot sink into concrete or tarmac!).

Get Active

For either the Derwent case study, or another case of flooding in the British Isles, make a note of:
- two physical causes of flooding
- two human causes of flooding
- two place names to remember
- two figures to remember.

Memorise them and ask a parent or friend to test you!

● Impacts of flooding

You need to:
- recognise that flooding has impacts on people and the environment.

Get Active

Complete the table below to summarise the impacts of flooding. For each impact write + or – next to it to show if it is positive or negative. Colour each statement, either green, if it impacts the environment, or red, if it impacts people.

Impacts	Positive or negative
Floodwater picks up pollutants like oil or chemicals and takes them downstream	
Spreads diseases	
People and animals can drown	
Provides water for crops	
Roads and railways may be disrupted	
Buildings and things inside them can be damaged	
Provides sediment which makes soil more fertile for growing crops	
Crops growing on the floodplain may be washed away	
Provides a habitat for fish which people can eat	
Once your house has been flooded, insurance becomes more expensive	

River management strategies

You need to be able to:

- know about hard and soft engineering strategies
- evaluate the management of a river outside the British Isles, with facts, figures and names of places.

There are two main responses to flooding: hard engineering and soft engineering. **Hard engineering** involves building large structures to try to control the river. **Soft engineering** means trying to reduce flooding without damaging the river for future generations.

Get Active

1 Match the 'methods' from column 1 with the correct 'purpose' from column 2 in the table below.
2 Decide whether each of the methods is an example of hard or soft engineering.
3 Think about why the hard engineering methods may not be sustainable in the long term.

Methods	Purpose
(a) Afforestation (planting trees)	(1) Water flows faster in a straighter and deeper river so it leaves the area without causing problems
(b) Building dams	(2) Make the river banks higher so more water can be held in the channel
(c) Building embankments or levees	(3) Trees take up water by their roots and reduce the amount that reaches the river, so it is less likely to flood
(d) Land use zoning	(4) Parts of the floodplain, often used as pasture in the summer, that can be allowed to flood in the winter – one form of flood storage area
(e) Flood walls	(5) Allow water to be stored rather than surging downstream and causing floods there. They also can be used for water supply, hydroelectricity and recreation
(f) Washlands	(6) Land that is at the highest risk of flooding is not used for housing but for playing fields or for pasture (from which animals are removed when the flood risk is high)
(g) Straightening and deepening the river	(7) Walls built beside rivers that are likely to flood in urban areas – they take up less space than levees

Case Study: A river management scheme – the Mississippi

The Mississippi is the fourth longest river in the world. It is an essential river for the USA, providing 18 million people with their water supply. Flooding is an almost annual event. Severe floods in 2001 caused $13 million of damage and 4400 people had to move.

Hard or soft engineering	Description	Evaluation – is management sustainable?
Hard engineering	Raised embankments called **levees** have been built 15 m high, for 3000 km along the river	Levees are partly blamed for the 2001 floods – they protect the area where they are built but push the problem downstream After a flood, silt is deposited on the channel bed instead of the floodplain, so normal water level rises to higher than the floodplain. Parts of New Orleans lie 4.3 m below river level increasing risk of flood damage
Hard engineering	Over 100 **dams** have been built on tributaries	Dams trap silt, preventing it reaching the delta (so birds like the heron are endangered there) or enriching farmland (so more fertiliser has to be used)
Hard engineering	Engineers cut through meanders to straighten 1750 km of channel, to make the river flow faster	A straightened river loses its variety of habitats for plants, fish and insects The river erodes the banks to resume its natural meandering course, so money and effort have been wasted
Soft engineering	**Afforestation** (tree planting) in the Tennessee Valley – trees absorb water	As well as preventing water from reaching the river, tree planting helps to reduce soil erosion and provides wildlife habitats and opportunities for recreation
Soft engineering	Safe flooding zones – houses near the river are bought and demolished (e.g. Rock Island, Illinois) and areas of floodplain are turned into green spaces	It is cheaper in the long term to prevent property damage than to compensate owners when the damage has happened Wetland habitats close to rivers can be preserved

Get Active

Imagine you are the mayor of a town flooded by the Mississippi. What would you say to residents who demand that more and higher levees are built to protect them in the future?

Sustainable management of coasts

● Human activity in the coastal zone

You need to:

- identify human activities in the coastal zone and how they conflict.

The coastal zone means the sea and the shore near the coast – so it includes lots of different types of area, including beaches, salt marshes, towns, islands and cliffs.

People use the coastal zone for many different things. Some of these are shown below in Figure 22 – you may be able to think of others!

Human activities

Residential – people living in towns and cities near the sea

Tourism – people go to the coast for holidays and days out – think of Portrush, with the beach and Barry's amusements, and the Giant's Causeway

Transport – roads and railways near the sea need protection to stop them getting washed away. Boats need harbours to keep them safe in storms

Industry – this includes drilling oil from under the sea, fishing, digging sediment up from the sea bed, building new ships and breaking up old ones

▲ **Figure 22** Human activity in the coastal zone

Sometimes, one use causes difficulty for other users. It can also get in the way of natural processes of erosion, longshore drift and deposition. This is called conflict.

Get Active

The following statements are examples of conflict in the coastal zone. For each one, decide which of the following are in conflict with each other: *residential, tourism, transport, industry, erosion, deposition*.

- **a** Mrs Currie used to have a great unspoilt view of the sea from her bedroom window. Now there is a big hotel in the middle of it, which she thinks is very ugly.
- **b** When Shania and Brad went on holiday to Limassol they were annoyed to see big noisy tankers moving in and out of the harbour.
- **c** Gill used to like driving down the tiny winding road to the rocky beach at Murlough Bay. One day she found the last stretch of the road was closed to traffic, and when she walked along it she found half the road had fallen into the sea.
- **d** The town of Dunwich used to be a thriving port, with about 3000 people living there. Now almost the whole village has fallen into the sea.
- **e** The town of Rye had an important harbour and had lots of industries. Hundreds of years ago the harbour started filling up with sand, and the town is now 2 miles away from the sea, because of all the sand which has built up.

● Coastal defences

You need to be able to:

- understand why coastal defences may be needed.

Coastal defences are ways of trying to stop the sea from eroding the land. People have been trying to do this for many years, because people use the coastal zone so much. It is predicted that sea level will rise as global warming continues. This means even more places will be affected by the sea. Some places could be flooded, and other places will have more storms which means more erosion by the waves.

Some people say this means we need more coastal defences. Other people say we should move away from the sea, and let the sea erode or deposit material naturally.

● Coastal management strategies

You need to:

- describe and explain coastal management strategies to keep the sea out and to protect the cliffs and beaches.

Coastal management means making decisions and taking action to control the land uses and natural processes happening at the coast. For example, if a building is at risk of falling into the sea because the sea erodes the land it is standing on, the council could decide to manage this by using coastal defences to try to stop the erosion.

1 *Strategies to keep the sea out.* Sea walls have been used for a long time to stop the sea. They look like a concrete wall at the back of a beach. Some sea walls are curved, others have steps, and others are straight. They are designed to stop the waves coming over the top and doing damage during storms. Curved walls are supposed to send the wave's energy back into the sea, and steps are supposed to help break up the wave's energy, so both are designed to stop erosion.

2 *Strategies to retain cliffs and beaches.* 'Retain cliffs and beaches' means keep them, rather than let them get eroded. Cliffs and beaches protect the land behind them from erosion. A wide beach can absorb the energy of the waves as they cross it. There are lots of ways of trying to protect cliffs and beaches.

   (a) Groynes – these are long fences made of heavy wood, stretching out from the beach into the sea. They are designed to trap sand moving along by longshore drift, and stop it moving away from the beach.

   (b) Gabions – these are wire boxes filled with stones, often put at the bottom of cliffs. They are designed to absorb and break up the waves' energy, and stop it eroding the cliff.

   (c) Beach nourishment – this is where large amounts of sand are brought onto the beach and deposited to help make a wider or deeper beach to protect the land behind it by absorbing the waves' energy.

Get Active

Match the diagrams with the following coastal management strategies as described on page 21.

- Sea walls
- Groynes
- Gambions
- Beach nourishment

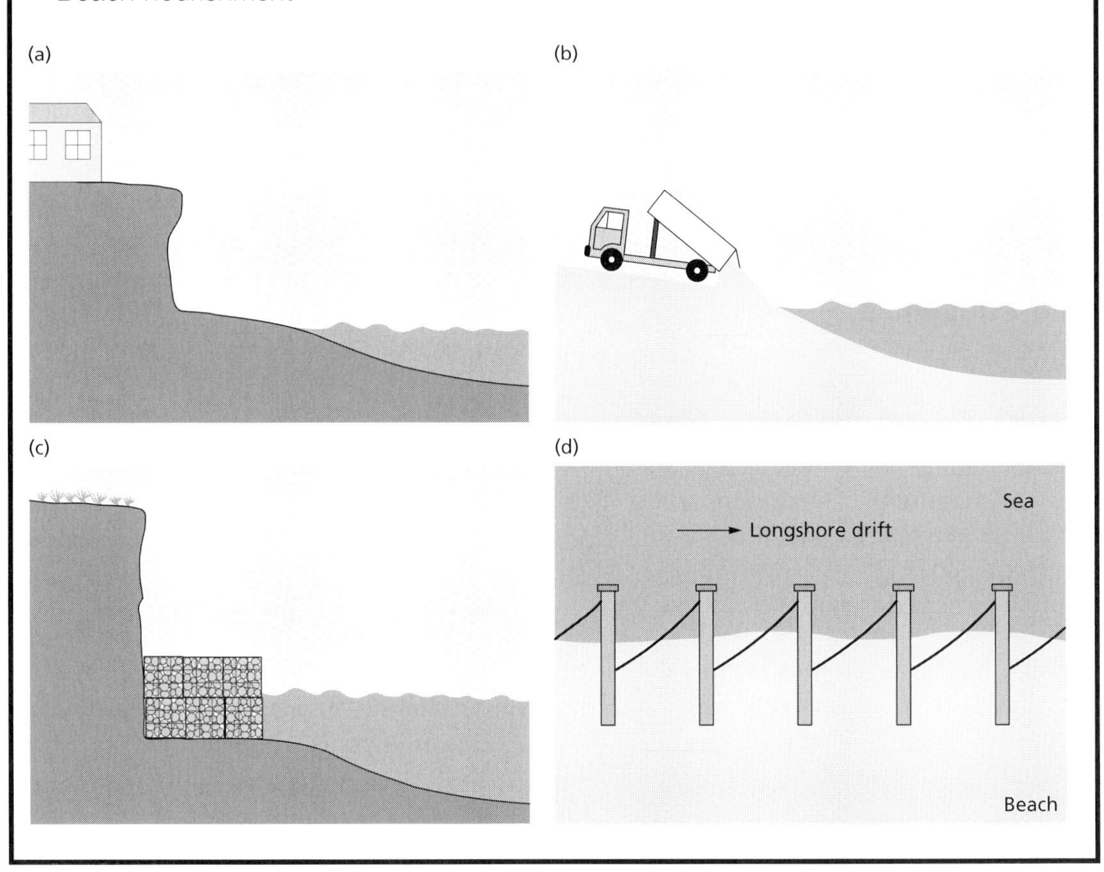

Case study: Coastal management in the British Isles – Newcastle, Co. Down

You need to:

- know about one place where the coast has been managed
- know what coastal management strategy has been used
- evaluate whether it is sustainable
- use place names and figures.

The table below shows the impacts of the strategies used in Newcastle. The impacts give evidence about how sustainable each strategy is, and this is summarised in the last column.

Strategy	Impact	Evidence for sustainability
1 Groynes – concrete groynes were built on the beach in the 1980s	(a) Trapped sand which was moving north-east	This is sustainable – it protected the beach for a while, resulting in more tourist enjoyment
	(b) Lasted about 20 years, then decayed and were useless	Unsustainable – the groynes do not last long, and eventually have to be replaced. This is expensive
2 Gabions – new wire mesh boxes filled with local stone were placed near the mouth of the River Shimna in 2006	(a) Protect recreation ground	Sustainable development – maintains an important tourist attraction
	(b) Have stabilised area of coast to allow pedestrians to use footbridge to access the rest of the promenade	Sustainable – gabions are able to break up wave energy better than walls or other methods, and seem to have been successful
	(c) Earlier gabions decayed over time and had to be replaced	Unsustainable – gabions need to be replaced after a time, which is expensive
3 Sea wall – a sea wall was originally built in the Victorian era – late 1800s – and a new one has been built since 2002	(a) The original wall protected buildings in Newcastle for many years	Sustainable development – original wall enabled the development of boarding houses and hotels, and protected them for a significant period of time
	(b) A storm in 2002 washed large parts of the wall away	Unsustainable – sea walls are not invincible and can be eroded. They are very expensive to replace – the new project cost £4 million
	(c) the new wall is curved so wave energy is directed back out to sea, instead of attacking the base of the wall	Unsustainable – if the beach is eroded it will not be able to protect the coast behind it. Fewer tourists will enjoy Newcastle if the beach is poor quality, which will bring less money to the town
	(d) reflected waves are increasing erosion of the beach on their way back down it	Unsustainable – any coastal management project has complicated impacts on the natural processes of erosion, longshore drift and deposition. It appears that the wall may have inadvertently caused erosion of the beach

Overall it appears that the coastal management strategies all have an element of unsustainability.

Get Active

For each strategy in the table on page 23, write down whether it was sustainable overall, and why. Make sure you include a place name, date or figure for each one.

You need to draw a line from each method on the left to the correct meaning on the right.

Specimen Exam Questions (Foundation tier)

1 (e) (ii) Match the following methods of coastal protection with their meaning: [3]

Groynes	Cages of stones used to protect the beach
Sea walls	Fences to stop sand moving along the beach
Beach nourishment	Costly structures used to protect important buildings
Gabions	Adding more sand or shingle to the beach

Higher Tier

1 (e) For a named case study within the British Isles, describe the coastal management strategies implemented and explain why these strategies were chosen. [8]

You should describe more than one strategy (groynes, gabions and/or sea wall at Newcastle, Co. Down) with **two** specific facts/figures, for example, groynes built on beach in **1980s**.

You need to explain how each strategy is meant to protect the coast, e.g. the groynes trap sand moving north-east and create a wider beach that absorbs wave energy.

Knowledge tests

Knowledge test I (Pages 1–2)

1 Name one input into the drainage basin cycle.
2 What term means that trees and other plants catch raindrops as they fall?
3 What term means water flowing slowly from rock into the river?
4 What is meant by the term percolation?
5 Is river discharge an input, transfer or output of the drainage basin cycle?
6 What term means water returns to the air from the drainage basin as vapour?
7 What is the difference between infiltration and throughflow?
8 What is the term used for a stream that flows into a river?
9 What is the term used for the point where two streams or rivers meet?
10 What does the term watershed mean?

Knowledge test II (Pages 3–15)

1 What happens to large angular rocks and pebbles as a river carries them downstream?
2 What erosion process is the action of water plus the load it is carrying?
3 What river transport process makes stones bounce or hop along the river bed?
4 What happens to a river's load when the river slows down?
5 Name the feature found at the bottom of a waterfall.
6 True or false? A river erodes on the outside bank of a meander bend because the water is deep and travelling fastest there.
7 What is deposited on a floodplain as a result of flooding?
8 Which type of waves have stronger backwash than swash?
9 What coastal transport process involves pebbles moving in a zigzag fashion along a beach?
10 In the list CLIFF, ARCH, CAVE, SPIT, STACK which coastal landform is the odd one out and why?

Knowledge test III (Pages 16–24)

1 State one physical cause of flooding of the River Derwent in 1999.
2 State one human cause of flooding of the River Derwent in 1999.
3 State one positive impact of flooding.
4 State two soft engineering strategies used to manage the River Mississippi.
5 State two hard engineering strategies used to manage the River Mississippi.
6 What is the connection between global warming and coastal defences?
7 How are sea walls shaped:
   a) to help break up wave energy
   b) to reflect wave energy back out to sea?
8 What term means adding sand or shingle to a beach?
9 What strategy is often used to retain (keep) sand or shingle on a coast where longshore drift operates?
10 Name two strategies, other than sea walls, used to protect the coast at Newcastle, Co. Down.

Our Changing Weather and Climate

Measuring the elements of the weather

You need to be able to:

- distinguish between weather and climate
- explain how elements of the weather are measured
- understand where to put weather instruments and why
- know where forecasters get their information.

Weather and climate

Weather is the day-to-day condition of the atmosphere, such as how hot it is, or whether it is raining. **Climate** is the average weather taken over about 35 years – the sort of weather we usually expect to get.

Get Active

Each of the sentences below refers to either weather or climate. Decide which is which.

a The day of the wedding turned out to be beautiful, with clear blue skies.
b Daisy's trampoline blew halfway across the garden in a storm.
c When she planned a picnic in November everyone said Brenda was crazy.
d In September, all the shops start to sell gloves, hats and scarves.
e Oonagh's mobile phone got ruined when she got caught in a rainstorm.
f The farmer in Spain had lots of irrigation to provide water for his crops, because he knew there wouldn't be enough rain.

● Measuring the elements of the weather

The elements of weather are the different things which make up weather. These are shown in Figure 1.

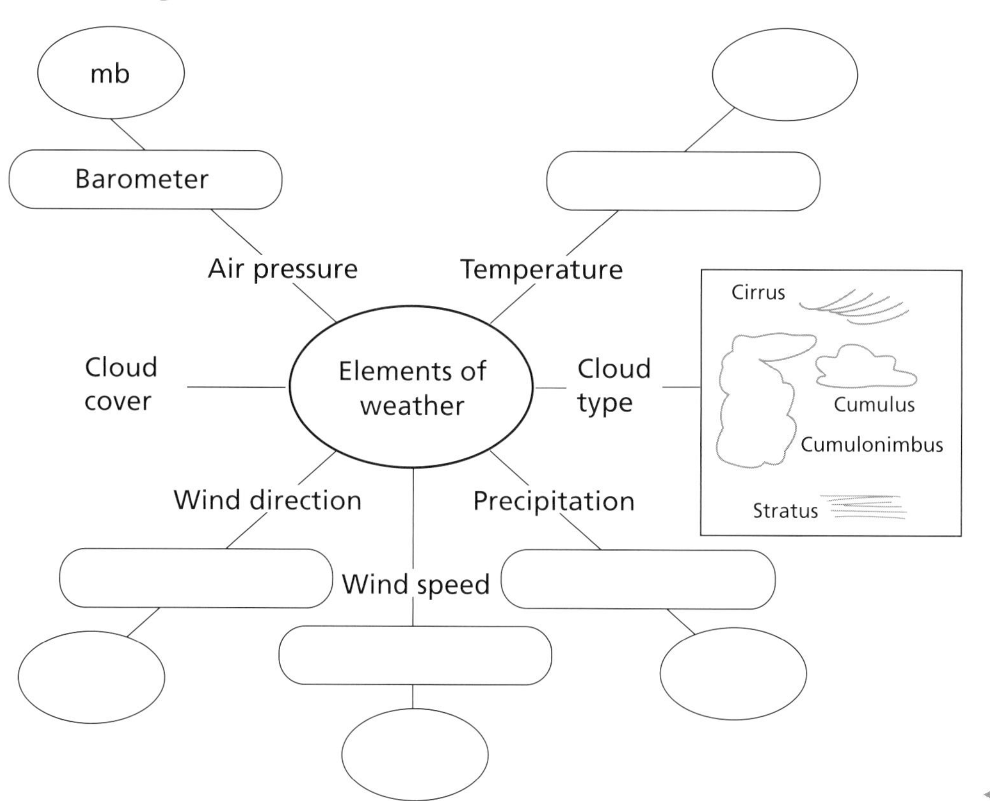

◄ Figure 1

Get Active

Copy and complete Figure 1 by selecting an instrument and appropriate units from the two lists. 'Pressure' has been completed for you.

Instruments
Rain gauge
Barometer
Wind vane
Maximum/minimum thermometer
Anemometer

Units
°C
Millimetres (mm)
Knots
Millibars (mb)
Eight points of the compass (N, NE, E, SE, S, SW, W, NW)

How to measure the weather

Precipitation	Read the level of water in the measuring cylinder of the rain gauge at the same time each day. Then empty it and reposition the rain gauge ready to record the rainfall over the next 24 hours
Temperature	A maximum and minimum thermometer has two sides, one for hottest and one for coldest temperatures. The mercury pushes metal markers up the tubes. The markers stay at their highest point even when the temperature changes. To find the temperature, read the bottom of each marker, then use the magnet to reset the markers to the top of the mercury
Wind speed	An anemometer usually has small cups which spin round as they catch the wind. Most anemometers have digital readouts which tell you the wind speed
Wind direction	On a wind vane, the arrow points to where the wind is coming **from**
Cloud cover	Look at the sky and estimate how many eighths (oktas) of the sky are covered in cloud
Air pressure	Read the needle on a barometer. There may be a second needle – this can be set to show what the pressure was at any one time, and the main needle then shows how pressure has changed

Where to put weather instruments

There are some instruments which need to be in special places to get accurate readings of the weather.

Rain gauge	This must be in the open, where nothing will stop the rain falling into it. It is often sunk partway into the ground, so it is less likely to be knocked over
Thermometer	We measure the temperature in the shade. This is always done so everyone understands what a temperature reading means. This means thermometers are usually kept in a white wooden box with vents in the sides which allow the air to pass through. This box is called a Stevenson screen
Wind vane	This must be high enough for nothing to shelter it from the wind, so that the true wind direction can be seen. A high roof or church spire are good places
Anemometer	This must be high up so nothing can shelter it from the wind

Get Active

Decide where each instrument should be in Figure 2.

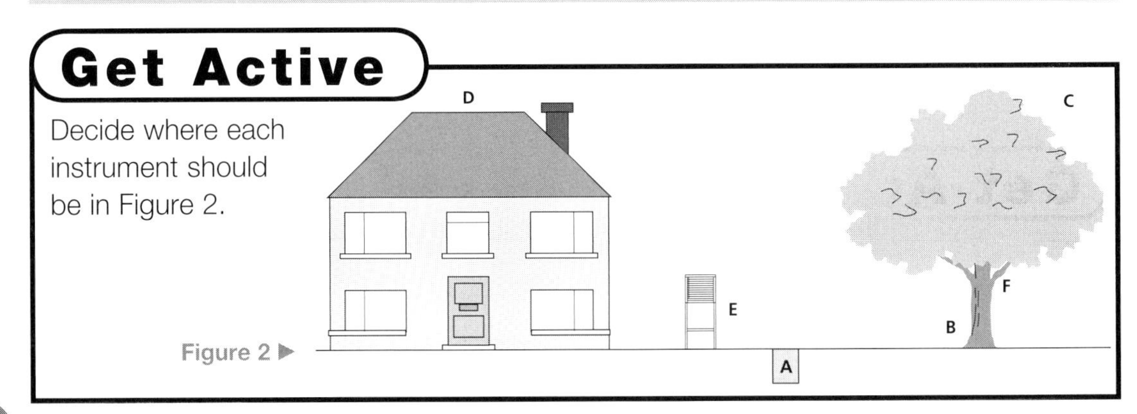

Figure 2 ▶

Sources of data for weather forecasting

Forecasters used to rely on pieces of seaweed or the colour of the sky to help them forecast the weather. Now they have a lot more data.

Figure 4 shows the sources of information used to make forecasts.

▲ Figure 3

Sources of information

Land-based weather stations – every 3 hours these record all the elements of the weather discussed in the last section. There are over 10,000 of these around the world.

Weather buoys – these are fixed or drifting in the sea, and use digital instruments to record the weather and transmit the information to a computer.

Ships – these have digital weather stations and record and transmit weather data.

Weather balloons – these are sent up through the atmosphere with digital instruments attached to measure temperature, pressure and wind speed. They transmit the information to a computer.

Satellites – these monitor the Earth from space and record data such as pictures of the clouds, and wind speed and direction.
Geostationary satellites stay above one place over the Earth all the time.
Polar satellites travel around the Earth about 14 times a day.

▲ Figure 4

Get Active

Try making up a short story using all these data sources to help you remember them. You could start with Florrie the forecaster going on a trip to the land-based weather stations, then getting caught up in the string of a weather balloon, accidentally landing on a satellite … the stupider the better as it will help you remember it!

Weather systems affecting the British Isles

You need to be able to:

- know the names of the four main air masses that affect the British Isles
- understand their temperature and moisture characteristics in winter and summer.

Air masses

An **air mass** is a large body of air (think of a giant 'blob' of air, perhaps the size of Western Europe) which takes on the temperature and moisture characteristics of the area where it is situated. If the air mass is stationary over the Sahara Desert, for example, then it becomes hot and dry. It then moves, as an air stream, and brings those characteristics with it. If it comes from the sea, it is called maritime and is moist. If it comes from a land area it is called continental and is always dry.

If it comes to the British Isles from the north it is cold and is called polar, but if it comes from the south it is warm and is called tropical. As the centres of continents get very hot in summer and very cold in winter, the temperature of a continental air mass differs a lot between summer and winter. Oceans only warm up slowly and hold their heat for a long time, so the temperature of a maritime air mass doesn't change much.

The four main air masses affecting the British Isles are shown in Figure 5.

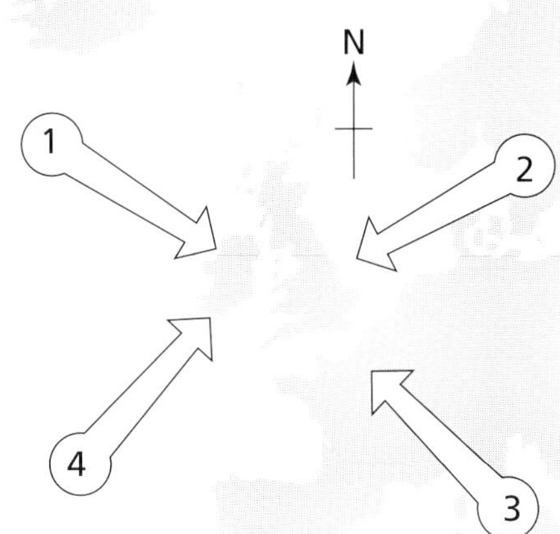

▲ Figure 5

	A Air mass name	B Temperature characteristics		C Moisture characteristics
		Summer	Winter	
1	Polar maritime	Warm/cool	Warm/cold	Wet/dry
2	Polar continental	Hot/cool	Warm/cold	Wet/dry
3	Tropical continental	Hot/cold	Mild*/cold	Wet/dry
4	Tropical maritime	Warm/cold	Mild*/cold	Wet/dry

* Mild means not very cold but not warm either.

Get Active

Copy and complete the table above by selecting the correct descriptions from the table in columns B and C.

Weather patterns associated with depressions and anticyclones

You need to be able to:

- know what weather occurs in different parts of a depression – at the warm and cold fronts in the warm sector
- know the sequence of change in the weather as a depression moves over the British Isles
- know what the weather is like in the British Isles during winter and summer anticyclones
- evaluate the effects (positive and negative) of depressions and anticyclones on the economy and people

When you write about this you need to be able to refer to places.

Anticyclones and **depressions** are weather systems which determine the weather of the British Isles and allow us to make weather forecasts.

	Anticyclone	Depression
Definition	An area of **high pressure**	An area of **low pressure** *(easy to remember – if you're depressed, you're feeling 'low')*
Air movement	Air is **sinking**	Air is **rising**
Cloud cover	**Clouds cannot form**	As it rises, air cools and condenses to form **clouds**
Wind speed	Winds are **gentle**, blowing **out** from the centre of high pressure. The isobars are far apart	Winds are **strong**, blowing **into** the centre of low pressure. The isobars are close together
Wind direction	Winds generally blow **clockwise** *(remember – **anti**cyclone and **anti**clockwise do not belong together)*	Winds generally blow **anticlockwise**
Duration	Anticyclones are slow-moving so weather **remains the same** for several days or even a week or more	A depression moves quickly so the whole depression can pass over a particular location in around **24 hours**
Weather	• Always **dry; little or no cloud; calm or gentle wind** • In **summer**, day temperatures are **high** (because of cloudless skies), night temperature can be **cool** as the Earth's heat radiates into the atmosphere • In winter, day temperature is **low** because days are short and sun is not powerful. At night **cloudless skies** allow the temperature to fall below 0°C and **frost** forms. Air near the ground is chilled and any moisture in it condenses to form **fog**	A **predictable sequence of weather** occurs as the depression moves over any location **1** First the warm front approaches. At the front warm air rises over cold air so clouds form and rain falls **2** Between the fronts is the warm sector, so temperatures rise a little, heavy rain stops and there may be some drizzle **3** The cold front arrives. Again the warm moist air is lifted over the cold air so clouds and heavy rain occur **4** Following the fronts, there is the cold sector, with lower temperatures. Cloud breaks up and showers become fewer

▲ **Figure 6** Comparison of anticyclones and depressions

Get Active

1 In the table on page 31, each row of information follows logically from the row above. Make sure that you can follow the logic and then draw out a simpler version of the table to show the key points, which are printed in **bold**. Remember that, in almost every way, the characteristics of a depression and an anticyclone are the opposite of each other.

2 The table below presents a list of ways in which anticyclones and depressions can have an impact on people. Copy the table and complete the final column by writing D beside the statements that are true for depressions, SA for summer anticyclones and WA for winter anticyclones. In the final column, put + if the effect is positive and – if it is negative.

Human impact	D, SA or WA	+ or –
1 Trees blown down, blocking roads		
2 Drought creates problems for farmers and gardeners		
3 Sequence of rain, showers and bright intervals ensure that crops have sufficient moisture		
4 Fog affects drivers on the M1 and causes delays for aircraft at Belfast International Airport		
5 Heavy rain may cause flooding, Cornwall 2009, for example		
6 Icy, slippery footpaths cause injuries to elderly pedestrians with increased admissions to the Royal Victoria Hospital		
7 TV weather forecasts advise the use of raincoats and umbrellas		
8 Lots of sunshine ripens crops in East Anglia		
9 Increased sales of ice cream and suntan lotion		

● Forecasting – synoptic charts and satellite images

You need to be able to:
- interpret synoptic charts and satellite images, distinguishing between depressions and anticyclones. A key is always provided with a synoptic chart so symbols do not need to be memorised.
- be able to predict what weather is likely to occur next.
- understand the limitations of forecasting.

Synoptic charts are weather maps that summarise the weather at a particular time. The date and time of day are clearly stated and should be noted as they help with the interpretation. A depression will have **fronts** (which are the boundaries between air masses) known as warm, cold and possibly occluded (see Figure 7) and the isobars are close together with the **lowest** value in the centre. Anticyclones have isobars that are further apart and the **highest** value is in the centre.

●●●●●	△△△△△	●△●△●△●△
Warm front	**Cold front**	**Occluded front**
Semi-circles like drawings of the sun reminding you of warmth. Always moves from west towards east	Shapes like jagged teeth reminding you of a cold 'biting' wind. Always moves from west towards east	This symbol is a mixture of both shapes. Found where warm and cold fronts meet

▲ **Figure 7** Symbols on a synoptic chart

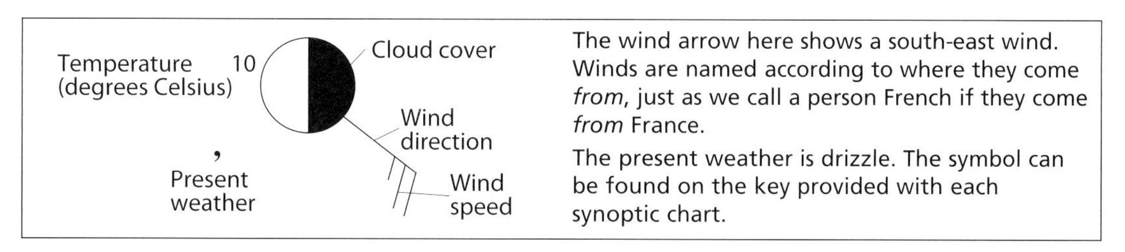

▲ **Figure 8** Weather station

Satellite images are photographs taken from space and sent back to Earth. On a satellite image a depression will show up as swirls of white cloud along the fronts, on a dark background. An anticyclone will be shown as clear skies, allowing the land and its coastline to be visible.

Weather forecasts and their limitations

A weather forecast is a prediction of the weather expected in an area. It may be for the next 24 hours (short range), the next 5 days (medium range) or the next 3 months (long range). A long-range forecast is less precise and dependable than the forecast for the next 24 hours.

Get Active

When the Met Office predicted a 'barbecue summer' for Britain in the summer of 2009 and was accused of getting it seriously wrong, it pointed out that Britain had enjoyed three months of temperatures slightly above the average which would have made for a very pleasant time were it not for the wettest July for 100 years!

1 Suggest two reasons why the Met Office attempts long-range forecasts when there is a strong possibility of getting it (partly) wrong.
2 Who do you agree with? Did the Met Office do their best, with some success, or did they mislead the public?

2006 Past Paper Exam Questions

1 **(b)** Study Figures 2a and 2b which show the weather systems over the British Isles on 15 and 16 September 2004.

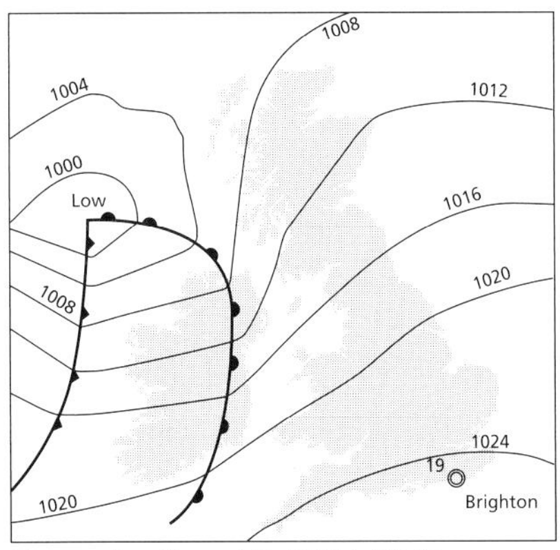

Figure 2a Sept 15, 2004 12 Noon

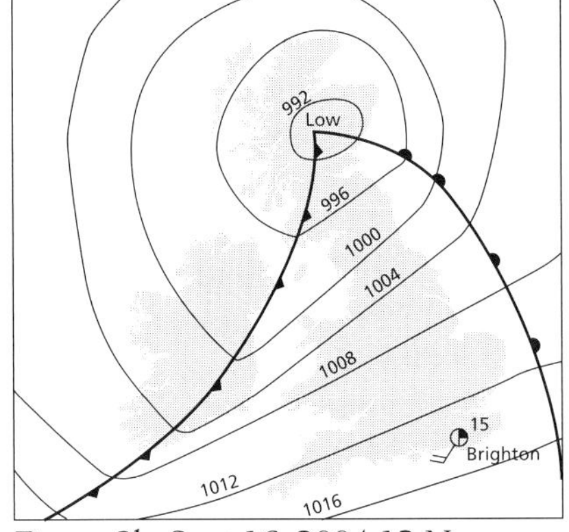

Figure 2b Sept 16, 2004 12 Noon

Key:

Cloud cover
oktas
○ 0
◑ 1
◔ 2
◕ 3
◐ 4
5
6
7
● 8
⊗ Sky obscured, e.g. by fog

Wind direction
Northerly
Easterly

Fronts
Warm
Cold

Wind speed
◎ Calm
1–2 knots
3–7 knots
8–12 knots
13–17 knots
18–22 knots

Weather conditions
● Rain
Drizzle
Mist
Fog
Thunderstorm

Past Paper Exam Questions

Answer the questions which follow.

Higher

(i) State the value of pressure over the south coast of England on 15 September (Figure 2a). [1]

(ii) 1 Name the front over Ireland on 15 September (Figure 2a). [1]

2 Describe how the location of this front changed from 15 September (Figure 2a) to 16 September (Figure 2b). [2]

(iii) The weather at Brighton on 15 September (Figure 2a) was warm and sunny. Explain why this was so. [3]

(iv) By 16 September (Figure 2b) the weather at Brighton had changed. Select two elements of the weather and explain why each changed. [6]

Foundation

(i) Underline the pressure value shown for the area near Brighton on 15 September (Figure 2a).

1016 mb **1020 mb** **1024 mb** [1]

(iii) Describe the weather at Brighton on 15 September (Figure 2a) by underlining the statement which completes each sentence below correctly.

1 Brighton is **in the warm sector/ahead of the warm front.**

2 It is **cold and wet/warm and dry** in Brighton.

3 **There is no wind/It is windy** at Brighton. [3]

There is a key for you to refer to here, so read it carefully. The answer is 'warm front'.

To get 2 marks you have to give enough detail. For example, 'front moved east across Britain to North Sea'.

This question follows the statement, consequence, elaboration pattern we discussed on page viii. Statement – there was no cloud (1). Statement and consequence – there was high pressure so no cloud formed so there was sunshine (2). Statement, consequence and elaboration – air was sinking, giving high pressure, so no cloud formed, so there was sunshine and temperatures went up (3).

Be careful with this – the isobar is not right on the south coast. Check for the direction the numbers go in – they are increasing towards the south, so it must be more than the closest isobar. The answer is 'Over 1024 mb' or 1024, 1025, 1026 or 1027 mb.

Choose two elements (from cloud cover, temperature, wind speed) and state how each has changed and explain why, for example, the clear skies on the 15th become cloudy by the 16th as Brighton is affected by a depression, where air rises, cools and condenses to produce cloud.

1024 – take the nearest option. Look carefully at the correct diagram!

There is no wind – again, the key gives you the meaning of the two circles inside each other.

It is warm and dry in Brighton – look at the synoptic symbol carefully and use the key.

Look carefully at the correct diagram! Brighton is ahead of the warm front.

The causes and consequences of climate change

● Greenhouse effect and global warming

You need to be able to:
- distinguish between the greenhouse effect and global warming.

Greenhouse effect

The **greenhouse effect** is needed for us to survive on the planet! The atmosphere around the Earth, especially some gases like carbon dioxide, acts like the glass in a greenhouse. It lets the sun's heat in, and stops some of it escaping. Without it we would freeze.

Global warming

This is exactly what it says – the globe warming up. There is evidence that world temperatures went up by 0.5°C last century. This may not sound like much, but even a small rise in overall temperatures could have huge impacts. Another term for **global warming** is climate change.

Why is climate change happening?

People argue about this – it is complicated because there are lots of different things happening.

- Natural climatic cycles
  - When the Earth goes round the sun it changes its orbit every 100,000 years. Sometimes it goes round the sun in a circle. At other times the path it takes is more like an oval, taking the Earth further from the sun at certain times of the year. When this happens, the Earth could cool down. When the orbit is circular the Earth could heat up.
  - Sunspots are tiny patches on the sun. Although the patches are slightly darker than usual, they are surrounded by brighter areas. Lots of sunspots means more brighter areas and more energy from the sun, so the Earth could get warmer. If there are fewer sunspots, it seems to make the Earth a little cooler.
- Volcanic activity – when volcanoes erupt they can send large amounts of ash, dust and gases into the atmosphere. This can cool the Earth down by shutting out some of the sun's energy.
- Human activity – people burn **fossil fuels** (coal, oil and gas, found underground and made from dead plants and animals over many years) which release carbon dioxide and other gases. These are greenhouse gases, and they add to the greenhouse effect, holding more of the sun's heat in and warming the Earth up. Most of these fuels are burnt in power stations and in motor vehicles.

Get Active

The following sentences all need either **greenhouse effect** or **global warming** in the gaps to complete them. Choose the right phrase.

a The gases round the Earth act like a blanket and keep the heat in. This is called the ____________________.

b The 10 warmest years recorded since 1860 have all been in the last 30 years. This is evidence of ____________________.

c Carbon dioxide is produced when people burn coal, oil and gas. This can make the ________________________ stronger, and lead to ____________________.

d It is thought the world could warm up by 5°C in the 21st century. This is ____________________.

e If the Earth stays close to the sun in its orbit, this could make it warmer. This is ______________________.

f If a volcano erupts it could send dust into the air, which would shade the Earth from the sun. This could slow down ____________________________.

g Without the ____________________ there would be no life on Earth because it would be too cold.

Effects of climate change

You need to be able to:

- evaluate the environmental, social and economic effects of climate change (positive and negative) for a case study
- use place names and other details.

Some of the main effects of climate change are shown in Figure 9.

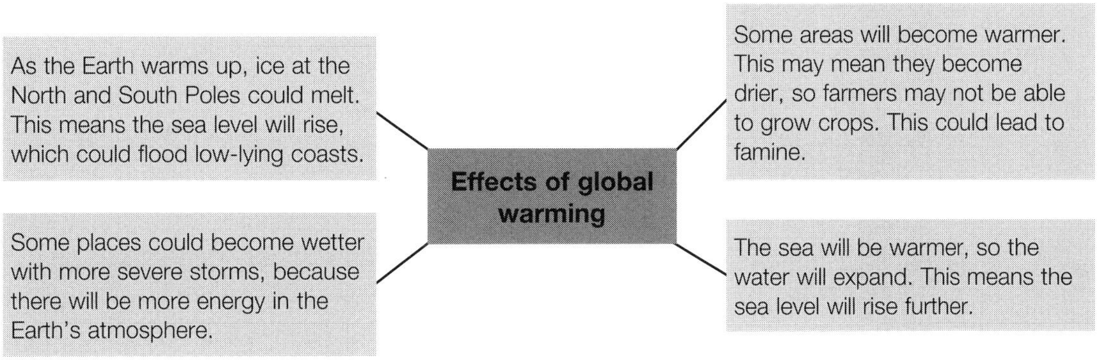

Effects of global warming

- As the Earth warms up, ice at the North and South Poles could melt. This means the sea level will rise, which could flood low-lying coasts.
- Some places could become wetter with more severe storms, because there will be more energy in the Earth's atmosphere.
- Some areas will become warmer. This may mean they become drier, so farmers may not be able to grow crops. This could lead to famine.
- The sea will be warmer, so the water will expand. This means the sea level will rise further.

▲ **Figure 9**

Get Active

The following statements are descriptions of effects that climate change has had or could have on the UK. For each, decide if it is positive or negative, and whether it is an effect on the environment, society or economy. Some may be more than one!

a Increases in pests like aphids and diseases in crops.
b Tropical diseases such as malaria.
c Rising sea levels could cause floods in low-lying areas such as near Strangford Lough, destroying wildlife habitats and farmland.
d Some animals could become extinct.
e Warmer air contains more moisture, so more rain – autumn 2000 was the wettest on record.
f Warmer winter temperatures mean less snow in Scotland, so skiing industry could disappear.
g Warmer summers in southern England – more tourism, such as £1.5 billion spent by tourists in July 1995.
h The south could become drier leading to water shortages.
i Farmers could grow more potatoes, tomatoes, vines and maize.
j Warm summers could lead to deaths in heatwaves – could kill as many as 3000 people.

Strategies to deal with climate change

You need to be able to:

- evaluate the sustainability of various strategies proposed to deal with climate change, referring to places to illustrate the points you make.

Strategies	Places	Details
International agreements	Kyoto Agreement 1997	Agreement signed by over 180 countries. MEDCs agreed to cut emissions of greenhouse gases to 5.2% below 1990 levels so that global warming can be slowed down.
	UN Climate Change Conference Copenhagen 2009	China, India and USA made some commitment to reduce their emissions.
Alternative sources of energy		Fossil fuels produce carbon dioxide and other pollutants when burned.
	Wind turbines work best in coastal or upland sites, for example, the coast of East Anglia and on top of the Antrim Plateau.	If we use nuclear power or renewable energy (wind, tidal or solar) then emissions of greenhouse gases will be reduced, but there are problems because the wind, tides and sun are not constantly available.
		Another alternative is to use biofuels such as landfill gas.
Reducing the use of private cars	A congestion charge of £8 per visit has been introduced in central London and has reduced traffic there by 15%.	Transport accounts for 35% of the energy used in the UK (more than industry, heating or lighting).
	In Belfast, Park & Ride schemes and bus lanes aim to help commuters to reach work more quickly by bus than by car.	A bus or train (public transport) carries many people using only slightly more fuel than a private car which often carries only one person.
		Reducing private and increasing public transport will reduce energy use and emissions.
Reducing deforestation	The President of Guyana wants rich countries to invest in sustainable development of rainforest resources (nuts, fruits, oils and medicines) rather than cutting it down for timber or to create grazing land.	Trees absorb carbon dioxide when growing but deforestation stops this and releases carbon dioxide when trees are burned.
		Preserving rain forests would slow down global warming.

Problems in securing international co-operation

Although we can each try to reduce our own carbon footprint, climate change is a *global* issue that cannot be solved by individual people or nations acting alone. International agreements are needed in order to make significant progress.

Agreement between nations is difficult to achieve because of the following issues:

- Individuals and rich nations are reluctant to make the major changes to their lifestyle that would be needed to cut carbon dioxide emissions significantly – people don't mind changing to low-energy lightbulbs, but most people don't want to give up driving a car!
- The MEDCs' factories produced lots of carbon dioxide while they were developing in the past. Now LEDCs want to catch up and want to be allowed to produce lots of carbon dioxide as they develop their own industries.
- Some countries such as the USA and Australia have been slow to recognise that climate change is a genuine problem. Even though the US President wants to cut emissions of greenhouse gases, other US politicians may oppose him.
- MEDCs have expensive technology which helps them reduce their carbon dioxide emissions. LEDCs can't afford this without help.

Get Active

If you could make a speech to the United Nations, what *one* strategy to limit global warming would you recommend, and why?

Knowledge tests

Knowledge test I (Pages 26–29)

1 True or false? Climate is the day-to-day condition of the atmosphere.
2 Name the instrument used to measure wind speed.
3 Name the instrument used to measure precipitation.
4 What are the units of measurement of pressure?
5 What are the units of measurement of temperature?
6 What element of weather is measured in oktas?
7 What is the white wooden box where thermometers are located called?
8 Why is this the most suitable place for measuring air temperature?
9 Where should a rain gauge be placed for accurate readings?
10 Name three possible sources of data used to create a weather forecast.

Knowledge test II (Pages 30–35)

1 What are the name and characteristics of the air mass that influences the British Isles when winds are blowing from the north-west?
2 Does an anticyclone have
a) high or low pressure, and
b) rising or sinking air?
3 Where is rain to be expected in a depression:
a) the cold sector,
b) the warm and cold fronts, or
c) the warm sector?
4 Which weather system, depression or anticyclone, brings strong winds?
5 Which weather system is responsible for fog and frost in winter?
6 What weather system, depression or anticyclone, is represented on a satellite image by a swirl of clouds on a dark background?
7 On a synoptic chart, what type of front has the symbol of a line of jagged 'teeth'?
8 State two impacts of anticyclones on the economy.
9 State one *positive* impact of depressions.
10 State two *negative* impacts of depressions.

Knowledge test III (Pages 36–40)

1. True or false? The greenhouse effect is the same as global warming.
2. True or false? Climate change results entirely from human activity.
3. What greenhouse gas is produced by burning fossil fuels?
4. State two causes of higher sea levels resulting from global warming.
5. State two positive effects climate change may have in the UK.
6. In which two cities have international conferences been held on climate change in:
   a) 1997 and
   b) 2009?
7. Name two alternative sources of energy we could use instead of fossil fuels.
8. What is meant by the term congestion charging?
9. How can LEDCs make money from their forests instead of allowing deforestation?
10. Why is it important to reduce deforestation?

The Restless Earth

Basic rock types

You need to be able to:
- understand how igneous, sedimentary and metamorphic rocks were made
- recognise what the rocks are like.

Formation of the basic rock types

Figure 1 shows the three basic rock types and how they were made.

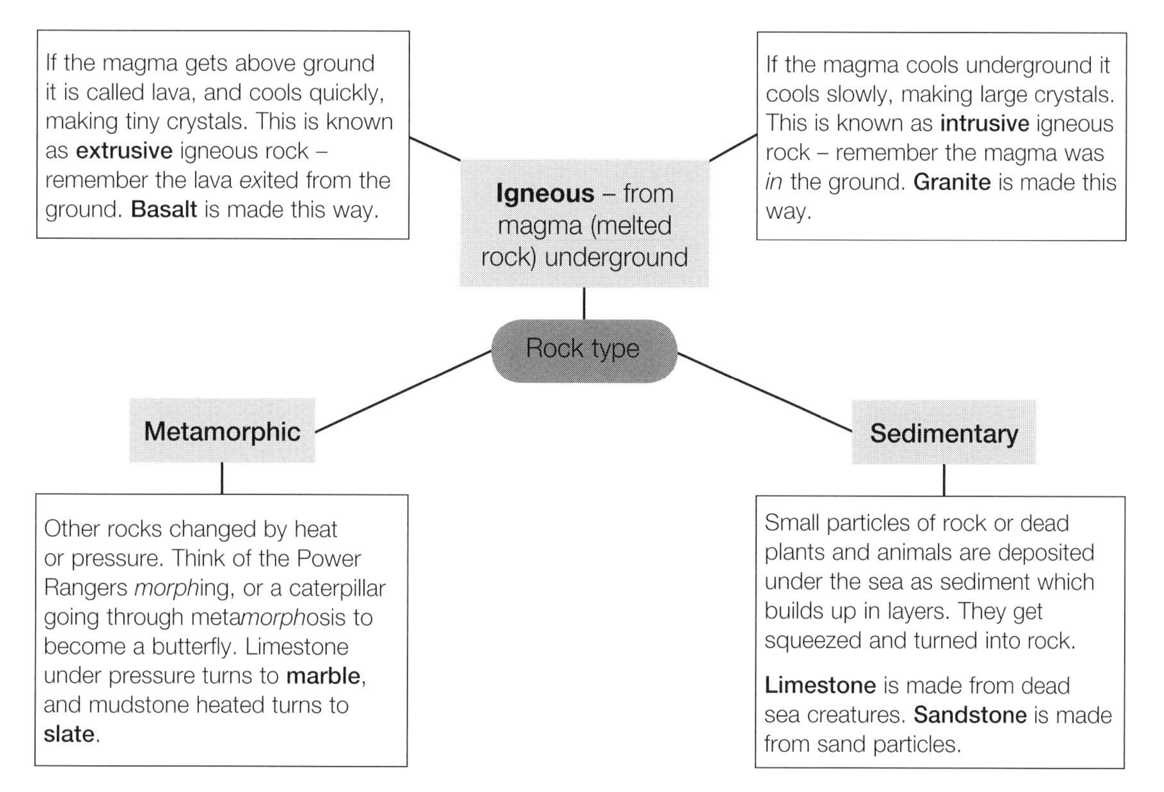

▲ Figure 1

● Characteristics of rocks

The table below shows the characteristics of some rocks. You need to be able to recognise the rocks from the characteristics, and describe the rocks.

Rock	Colour	Other features
Basalt	Dark grey/black	Glittery speckles
Granite	Speckled grey, white, black, pink	Very hard, large crystals visible
Limestone	Grey, white or yellow	May have fossils, fizzes when a drop of acid is added
Sandstone	Yellow/orange	Often see grains of sand, may rub off
Slate	Dark grey	Layers split apart easily, smooth, can be marked
Marble	White, swirls of colour	Can be highly polished for fireplaces and floors

Get Active

1 Get someone to read out the colour and other features for each rock. Try to name the rock.

2 Figure 2 below shows how different rock types are formed. Decide which rock type belongs in each space, choosing from the words in bold in Figure 1.

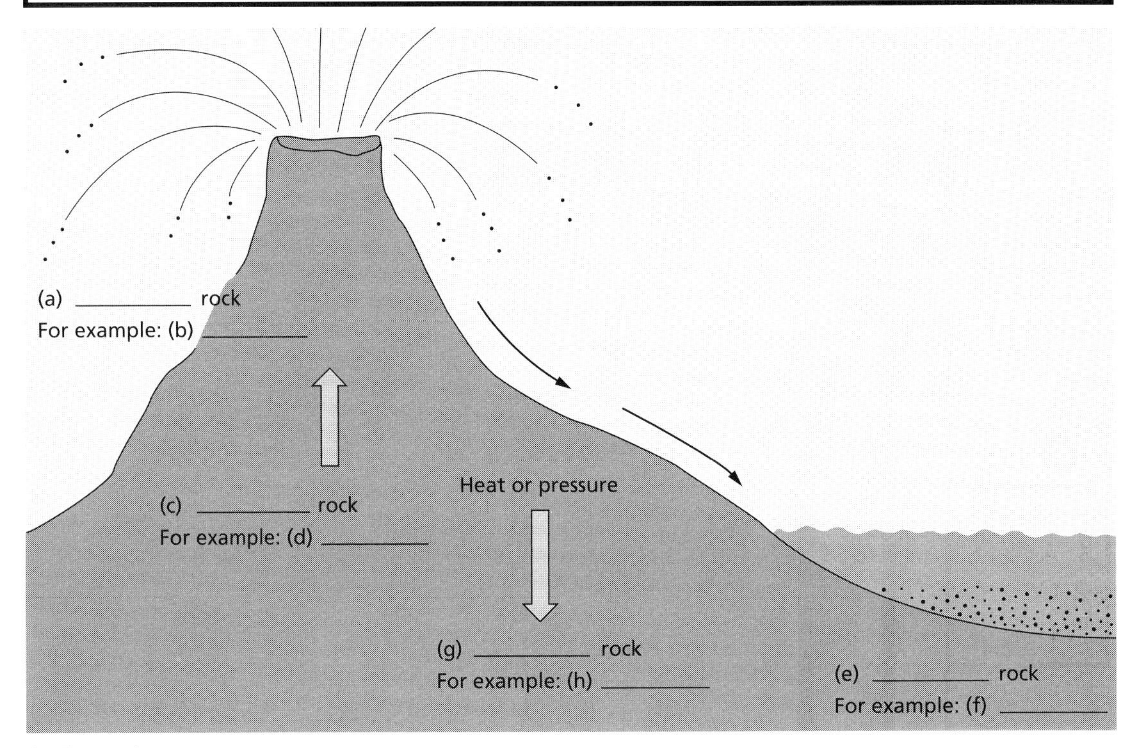

▲ Figure 2

Plate tectonics theory

You need to be able to:

- describe the structure of the Earth – core, mantle and crust
- know that the crust is made up of a number of plates
- understand how convection currents cause plate movement.

Structure of the Earth

The Earth's structure can be divided into **core**, **mantle** and **crust**, as shown in Figure 3. The crust is not continuous and smooth, like the skin of an apple, but is made up of segments rather like the hexagons that make up a football. The segments of crust are known as **plates** and they are able to float on the semi-liquid mantle below. On average, plates move about 7 cm a year. The intensely hot core heats the mantle and this causes **convection currents**. As these currents reach the top of the mantle and spread out, they drag the plates slowly apart (see plates B and C in Figure 3). When the convection currents cool and begin to sink back towards the core, they drag the plates above them closer together (see plates A and B in Figure 3).

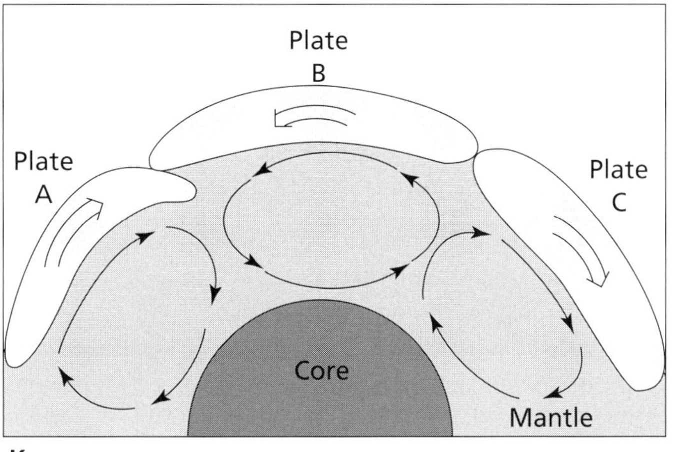

▲ **Figure 3** Convection currents in the mantle

Get Active

From the glossary on pages 116–120 learn the definitions of:

1 A plate.

2 Convection currents.

Repeat them aloud to a pet, a parent or other member of your family.

Types of plate margin

You need to be able to:

- understand the processes that occur at different plate margins
- explain what landforms are found at different plate margins:
  - constructive: mid-ocean ridge
  - destructive: subduction zone and ocean trench
  - collision zone: fold mountains
  - conservative: fault lines.

The edge of a plate, where it meets another, is called a plate margin or boundary. A map of the world's plates showing destructive, constructive and conservative plate boundaries and the global pattern of earthquakes and volcanoes is found on page 84 of your main textbook, *Geography for CCEA GCSE Second Edition*. The 'ring of fire' is the name of the zone of active volcanoes found along the edge of the Pacific plate.

What happens at different types of plate margin?

	Destructive plate margin	Constructive plate margin	Conservative plate margin
Plate movement	Plates come together	Plates move apart	Plates slide past each other
Explanation of name	Crust is destroyed	New crust is formed or constructed	Crust is conserved – neither destroyed nor added to
Processes	When plates meet, the oceanic plate is forced to bend and go down into the mantle beneath the other plate. This is the process of **subduction** and it triggers earthquakes as the plate bends. Friction and the heat of the mantle melt the descending plate, forming magma. This magma rises and forms volcanoes on the continental plate	As the two plates move apart, molten rock or magma rises from the mantle to fill the gap, forming new crust	Two plates slide past one another, along a **fault** line. Friction between them means that they tend to stick until pressure builds up and is released in a sudden jerking movement: an earthquake
Features	Earthquakes, volcanoes, fold mountains and an **ocean trench**	Earthquakes, volcanoes and **mid-ocean ridge**	Earthquakes but no volcanoes

(a) Destructive plate boundary

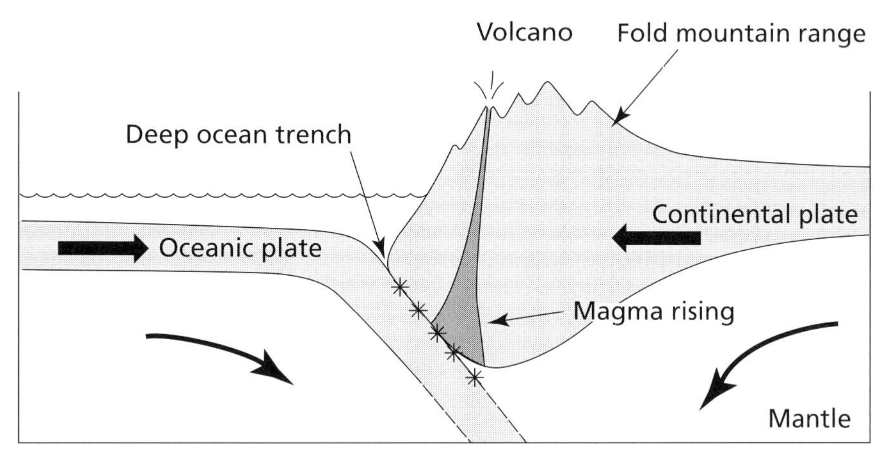

(b) Constructive plate boundary

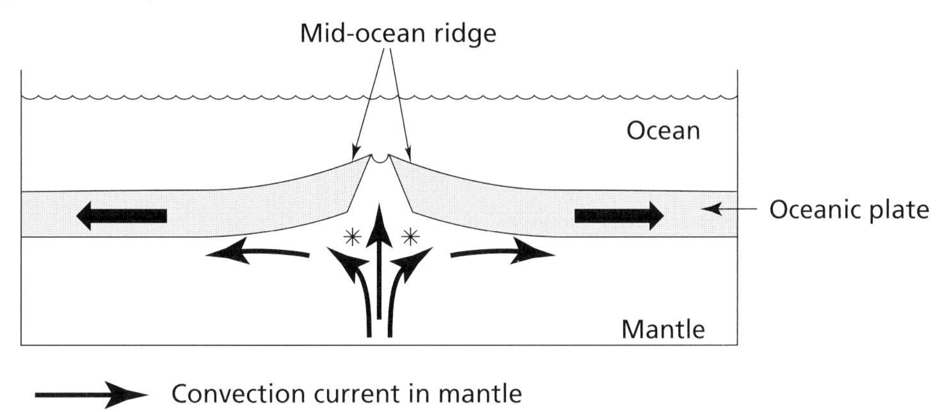

(c) Conservative plate boundary

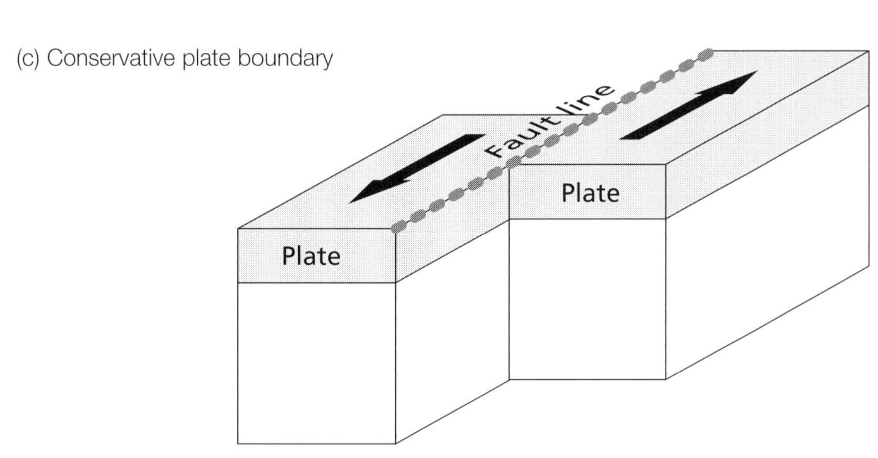

▲ **Figure 4** Three types of plate boundary

Collision zones

When two continental plates meet there is a head-on collision. A long time ago, India was separated from the rest of Asia by a sea. As the Indo-Australian plate crashed into the Eurasian plate, the sediments on the seafloor between India and Asia were folded and lifted upwards, making fold mountains called the Himalayas. Sometimes magma starts to rise through the mountains but cools before it reaches the surface, making intrusive igneous rock underground. Earthquake activity is common because so much rock movement is going on.

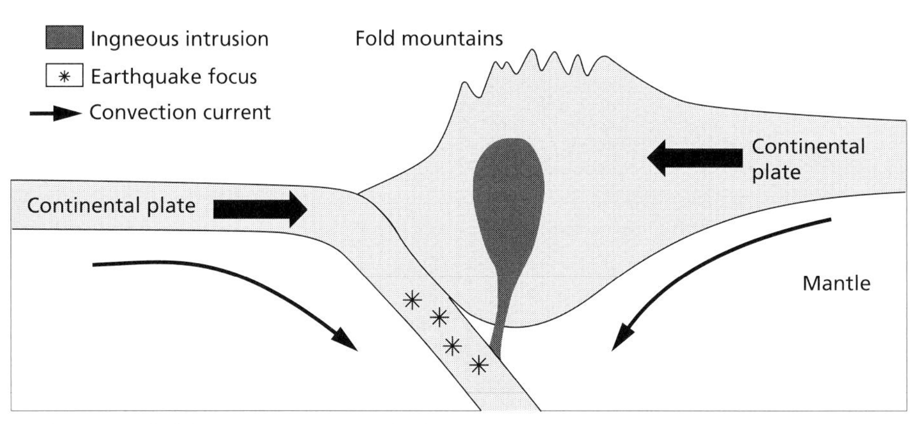

▲ **Figure 5** Collision zone (cross-section)

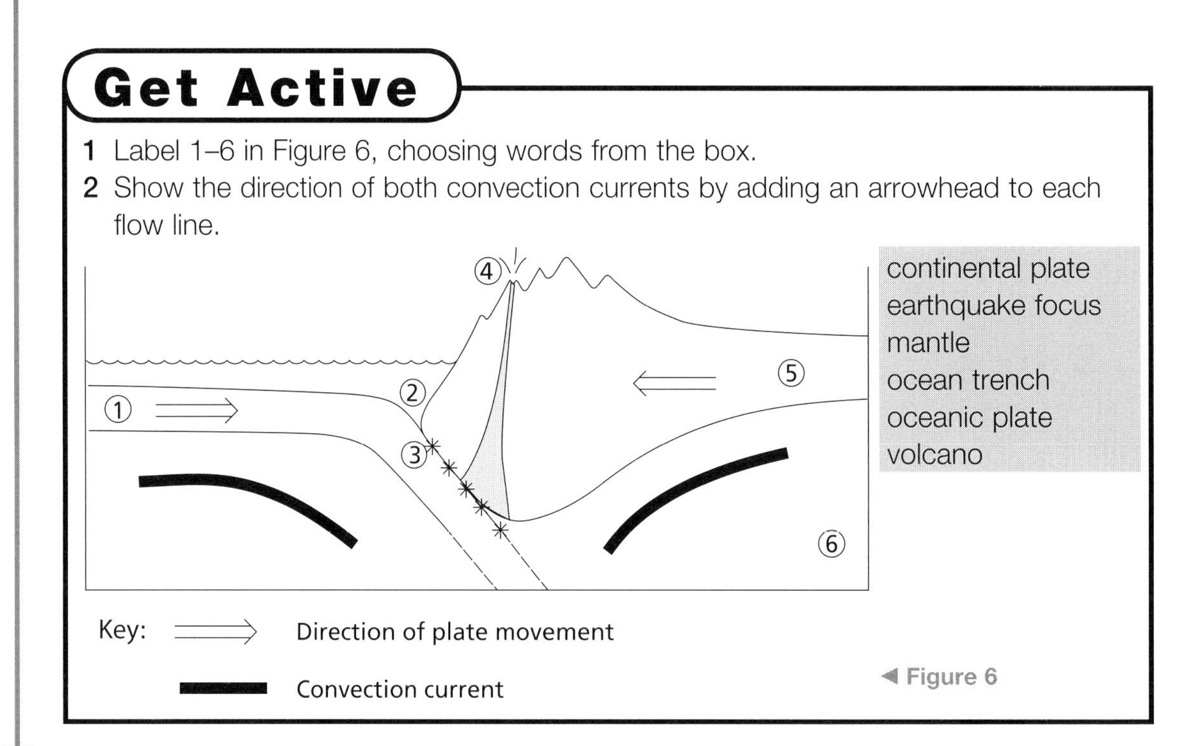

Get Active

1 Label 1–6 in Figure 6, choosing words from the box.
2 Show the direction of both convection currents by adding an arrowhead to each flow line.

◀ **Figure 6**

Tectonic activity in the British Isles

You need to be able to:

- identify a lava plateau, basalt columns and volcanic plugs
- explain how they were formed.

Landscape features created by tectonic activity

Lava plateau

- This is a large upland flat area made when runny lava erupted from cracks called fissures and flowed a long way before cooling down.
- It spread out, filled in dips and made a flat surface.
- Several eruptions happened over a long time period. This gave layers of basalt on top of each other.
- The Antrim plateau, stretching from Belfast to the North Antrim Coast, is a lava plateau.

Basaltic columns

- These are tall thin blocks of basalt.
- They have a variety of shapes – some are hexagonal, but not all. Usually when basalt cools, it cools quite quickly.
- However, if the lava fills a dip in the ground, it makes a deep pool and cools more slowly than usual. As it cools the lava shrinks, and pulls inwards, making cracks. These are the edges of the basalt columns.
- The most famous basalt columns are at the Giant's Causeway. Go on a virtual visit at www.virtualvisit-northernireland.com and type Giant's Causeway into the keywords box.

Volcanic plugs

- These are hills which rise steeply from the surrounding land.
- They were made when magma in a volcano cooled down inside the vent, and turned to rock called dolerite. This is harder than the surrounding rock, so it is left sticking up when the other rock is worn away.
- Slemish in Co. Antrim is a volcanic plug.

An earthquake in the British Isles

You need to be able to:
- describe and explain the causes and impact of one earthquake in the British Isles
- use place names and figures.

Case study: Market Rasen earthquake, 2008

Causes
- Many rocks in the British Isles have faults, or cracks, in them. The forces that make plates move can be transferred even to the middle of the plates, and put pressure on faults. Under this pressure the rocks can move. This makes shock waves which we feel as an earthquake.
- At Market Rasen in Lincolnshire, the earthquake happened because rocks on a fault line about 20 km underground moved suddenly. The earthquake was 5.2 on the Richter scale – strong for the British Isles.

Impacts
- Many people were woken by a noise and shaking.
- It was felt as far away as Northern Ireland and London.
- Some buildings were damaged, mostly cracks and chimneys broken. Some roads (such as Trinity Street in Gainsborough) were closed for a while to make sure nothing fell on cars or people below.
- The church in Market Rasen had £10,000 damage done when the stone cross fell and hit the roof.
- A man in Yorkshire had a fractured pelvis when his chimney fell on top of him in his bed.
- There were power cuts in some areas.
- More than 5000 people phoned the police in an hour.

Get Active

Draw a spider diagram to show the causes and impact of the Market Rasen earthquake. Use blue for the causes of the earthquake and red for the impacts. Make sure you include three place names or numbers.

Earthquakes: can they be managed?

- An **earthquake** is a shock, or series of shocks caused by a sudden Earth movement.
- The **focus** is the point in the Earth's crust where the earthquake occurs and the **epicentre** is the point on the Earth's surface directly above the focus, where its effects are felt first.
- An instrument called a **seismograph** records shock waves, and the magnitude of the earthquake is measured on the **Richter scale**.

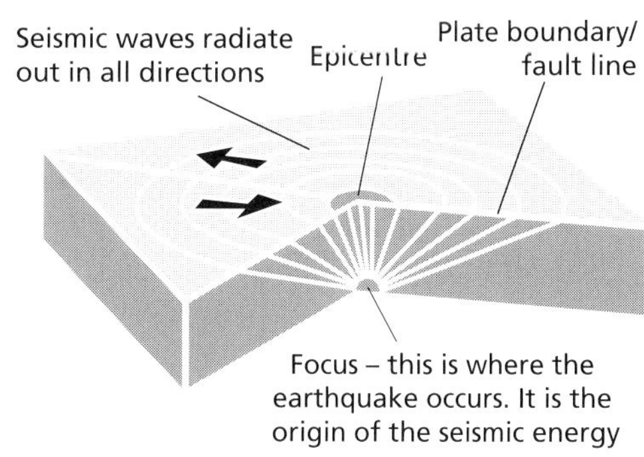

▲ **Figure 7** Features of an earthquake

Global distribution

You need to be able to:
- understand the global distribution and causes of earthquakes, in relation to plate boundaries.

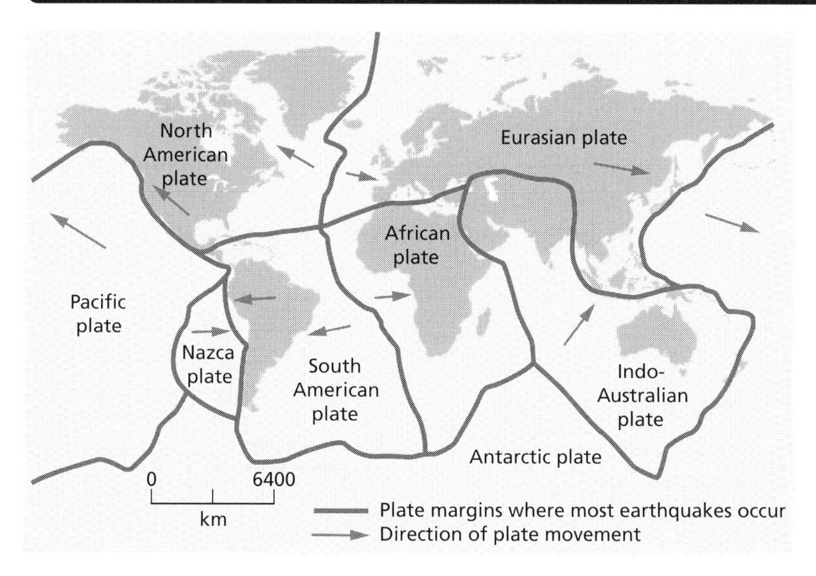

◀ **Figure 8** Global distribution of earthquakes

Figure 8 shows that the distribution of earthquakes is linear (in lines or zones) and closely related to plate margins. They are mainly found:

- around the Pacific Ocean (part of the ring of fire)
- roughly in a north–south line in the middle of the Atlantic Ocean
- roughly in an east–west line across southern Europe, the Himalayas and South-east Asia.

Earthquakes happen at plate margins because this is where friction occurs and stresses build up as sections of crust move past each other.

Physical consequences of earthquakes

You need to be able to:

- understand how liquefaction and tsunamis are caused by earthquakes.

- **Liquefaction** happens when rock or soil containing water is shaken by an earthquake. The water rises to the surface and turns the soil to liquid mud. Any buildings resting on this sink into the mud and collapse.
- A **tsunami** is a large wave of seawater, caused by an earthquake under the sea. The shock waves from the plate movement create small waves (about 30 cm high) on the ocean surface. As they move towards the shore, and come into shallower water, they become much larger – up to 30 m high. Before a tsunami, people on the shore observe a dramatic fall in sea level, exposing areas normally under water, and this can act as a warning.

Get Active

Fill a small cereal bowl with sand and add enough water to make it damp. Place three or four Lego bricks on the sand to represent a building. Set the bowl on a table and, using both hands, tap opposite sides of the bowl repeatedly and watch what happens to (1) the sand surface and (2) the Lego building.

Case studies: Earthquakes in an MEDC and an LEDC

You need to be able to:

- know what caused each earthquake, including the names of the plates involved
- describe the short- and long-term impacts on people and the environment
- evaluate the management response to earthquakes – prediction, precautions and strategies (immediate and long term) implemented after the events.

Case study: An earthquake event in an MEDC – Kobe, Japan

- Date: 1995
- Magnitude: 7.2 on the Richter scale
- Causes: Japan is on a destructive plate boundary where two oceanic plates (the Pacific plate and the Philippine plate) are dipping underneath the Eurasian plate. The two oceanic plates experience friction as they bend and descend into the mantle, so pressure builds up. When the pressure is finally released it creates a severe earthquake.

Impacts

	Short term	Long term
Impacts on people	• 200,000 buildings collapsed • 5500 people died • A 1 km stretch of the Hanshin Expressway collapsed • 80% of quays in Kobe docks were destroyed • More than 150 fires started by broken gas mains • 250,000 people made homeless, many camping outdoors in night temperatures of –2°C	• Water, gas and electricity not fully restored until 6 months later • Hanshin Expressway closed for over a year • Homeless people living in temporary accommodation • Unemployment resulting from the closure of Mitsubishi and Panasonic factories
Impacts on the environment	• **Liquefaction** – water in clay rock below parts of Kobe rose up turning it into mud and causing collapse of buildings • The ground moved 1.2 m vertically and 2.1 m horizontally	• Most commercial buildings were repaired or replaced to higher safety standards

Management responses

Prediction and/or precautions before the event	Immediate and long-term strategies after the event
As earthquakes are common in Japan, precautions are taken: • buildings are designed to withstand earthquakes, with springs or rubber pads to absorb the shock waves • earthquake drills every year, on the anniversary of the disastrous 1923 Tokyo quake, allow emergency teams and individuals to practise and improve their responses • people keep earthquake kits (including blankets, bottled water, rice and a radio) in their homes	Immediate attempts by fire crews from Kobe and nearby Osaka and Kyoto to put out the fires. Emergency shelter and food provided immediately from elsewhere in Japan. In the longer term there have been: • increased spending on research into earthquake prediction, for example, by satellite detection of minute distortions in the Earth's crust • stricter building regulations such as flexible steel frames for high-rise buildings and fire-resistant materials for houses

Get Active

Draw a spider diagram to sum up the Kobe earthquake: its causes, impacts and management responses. If you have a different earthquake case study, draw a diagram for it instead, making sure that you include all the headings from the Kobe study.

Case study: An earthquake event in an LEDC – Indian Ocean

- Date: Boxing Day, 2004
- Magnitude: 9.2 – the second largest ever recorded
- Causes: A fault line exists where the Indo-Australian plate is pulled (subducted) under the Sunda plate which lies on the southern edge of the Eurasian plate. A sudden slip of 15 m occurred along a 1600 km section of this fault, lifting the seafloor several metres and triggering a **tsunami**.

Impacts

	Short term	Long term
Impacts on people	• 66% of Sri Lanka's fishing boats destroyed • At least 125,000 people injured • 1.1 million people left homeless when tsunami devastated coastal towns and villages	• 187,000 people confirmed dead, with 43,000 missing • 17 low-lying Maldive Islands had their freshwater supplies contaminated by seawater, making the islands uninhabitable for decades • Emotional distress of losing relatives, especially when no body was found for burial • Rebel group GAM declared ceasefire with the Indonesian government (to allow relief and reconstruction) and signed a peace agreement in 2005
Impacts on the environment	• A tsunami 30 m high travelled from epicentre to countries all around Indian Ocean	• Damage to coral and mangrove ecosystems on the coasts • Worldwide rise in sea level of 0.1 mm as water spilled out from the Indian Ocean after its seafloor was raised

Management responses

Prediction and/or precautions before the event	Immediate and long-term strategies after the event
• No official warning system in place • A 10-year-old British girl on holiday in Phuket, Thailand saw the water receding and warned her parents, so the beach was evacuated saving about 100 lives	• International aid and expertise from around the world – over US$7 billion • State of emergency declared in Sri Lanka, Indonesia and the Maldives to allow orderly distribution of aid • Indian Ocean Tsunami Warning System set up by UNESCO with 25 new seismographic stations to detect future tsunamis and provide warnings for countries in the region

Get Active

Make a copy of all the bullet points in the previous two tables. Jumble them up and then practise putting them under the correct heading: impacts or management response, short term or long term, people or environment.

Name an earthquake you have studied. There is no mark for this but you do have to be specific to one earthquake to get top marks. This is a case study question, so you should be aiming to give details like place names and numbers.

2006 Past Paper Exam Questions (Higher Tier)

(e) Name an earthquake you have studied.

1 Explain the cause of the earthquake. [3]

2 Describe **two** strategies put in place after the event to reduce loss of life in future earthquakes. [6]

For 3 marks you need to give detail, including one fact or figure such as naming the plates. For example, 'Japan is on a destructive plate boundary where two oceanic plates (the Pacific plate and the Philippine plate) are dipping underneath the Eurasian plate. The two oceanic plates experience friction as they bend, so pressure builds up. When the pressure is finally released it creates a severe earthquake.'

The marks break down into 3 for each strategy – each one needs statement (S), consequence (C) and elaboration (E).
For example, 'they introduced stricter building regulations (S) so that buildings would be less likely to collapse and kill people (C). For example, high-rise buildings had to have flexible steel frames (E).
They increased spending on research into earthquake prediction (S), hoping to be able to give people warning to leave their homes if an earthquake was about to happen, so that they would be less likely to be killed (C). For example, they used satellites to detect tiny changes in the shape of the Earth (E).'

Knowledge tests

Knowledge test I (Pages 43–48)

1 Which category of rocks, igneous, sedimentary or metamorphic, forms from sediment that builds up in layers?
2 Which igneous rock, basalt or granite, forms from magma cooling slowly underground?
3 Name two examples of sedimentary rock.
4 Name a metamorphic rock that forms:
a) under pressure
b) by extreme heat.
5 What causes the plates of the Earth's crust to move?
6 If plates move apart, is the plate margin constructive or destructive?
7 At what type of plate margin is an ocean trench formed?
8 At what type of plate margin is a mid-ocean ridge formed?
9 What term means the process of one plate bending under another and being forced down into the mantle?
10 What plate movement happens at a conservative margin?

Knowledge test II (Pages 49–50)

1 Where in the British Isles would you find a lava plateau?
2 True or false? A lava plateau forms from a single eruption of runny lava from a volcanic cone.
3 True or false? The basalt columns at the Giant's Causeway were formed when volcanic lava cooled faster than usual.
4 True or false? Slemish is a volcanic plug formed when magma cooled in the vent of a volcano.
5 In what year did Market Rasen experience an earthquake?
6 How strong was the earthquake on the Richter scale?
7 The Market Rasen earthquake was caused by:
a) volcanic activity,
b) movement of rocks on a fault line or
c) coal mining?
8 Give two examples of how far away the effects of the Market Rasen earthquake were felt.
9 How many people were:
a) killed and
b) injured by the earthquake?
10 Which building in Market Rasen was most seriously damaged?

Knowledge test III (Pages 51–55)

1 Name the point in the Earth's crust where an earthquake occurs.
2 Name the point on the Earth's surface directly above.
3 Name the process where water rises to the ground surface during an earthquake, causing buildings to collapse.
4 Name the plates involved in the 1995 Kobe earthquake in Japan.
5 How many people were:
a) killed and
b) left homeless by the Kobe earthquake?
6 What takes place every year in Japan on the anniversary of the disastrous 1923 Toyko earthquake?
7 Name the plates involved in the 2004 Boxing Day earthquake in the Indian Ocean.
8 What was the magnitude of this earthquake?
9 What was the cause of thousands of deaths from this earthquake all around the Indian Ocean?
10 Approximately how many people were left homeless by it?

A People and Where They Live

Population growth, change and structure

You need to be able to:

- know what is meant by birth rate and death rate
- know how and why birth rates, death rates and the population of the world have changed since 1700.

● World population growth

Birth rate is:

- the number of live births
- per 1000 people
- per year.

Death rate is:

- the number of deaths
- per 1000 people
- per year.

Try to remember all three parts of these definitions!

When the birth rate is more than the death rate, that means more people being born than dying in any year. This means the population will increase. The amount it increases by is called the **natural increase**. To work it out, take the death rate away from the birth rate.

Example:
Birth rate = 30
Death rate = 20
Natural increase = 30 – 20 = 10.

The natural increase is 10 per 1000, or 1% growth rate (obtained by dividing by 10).

When the death rate is more than the birth rate, more people die than are born in any year. The population will decrease.

The amount it decreases by can be worked out in the same way as above.

Example:
Birth rate = 20
Death rate = 25
Natural decrease = 20 – 25 = –5.

The natural decrease is –5 per 1000, or –0.5% growth rate.

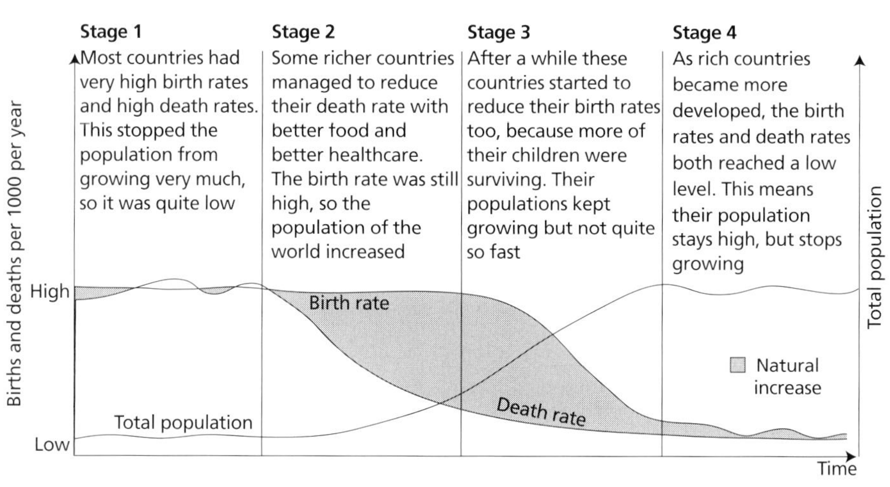

Figure 1 Changes in birth rate, death rate and world population since 1700

The population of the whole world has been affected by the changes shown in Figure 1.

In more developed countries like the UK and USA, the populations have already increased and now stay more or less the same.

Some less developed countries like India and Mexico are now in the middle stages shown in Figure 1, and their populations are growing quickly.

This means the world population is going to grow a lot more in the next few years.

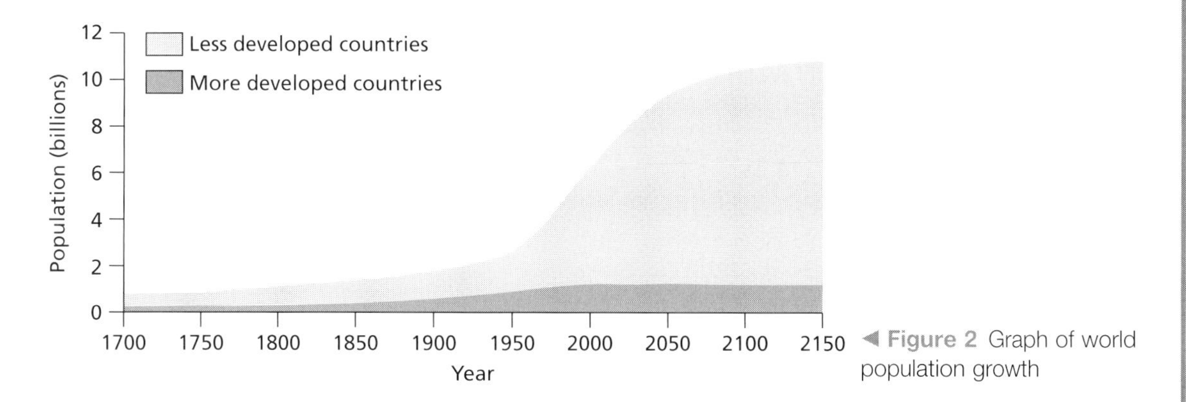

Figure 2 Graph of world population growth

Get Active

Decide which statements are true and which are false:

1. The population of the world is decreasing.
2. The population of the world is increasing.
3. World population grew slowly until 1950.
4. World population grew rapidly from 1700.
5. World population grew rapidly from 1950.
6. The population in 2000 was growing fastest in MEDCs.
7. The population in LEDCs is growing much faster than in MEDCs.
8. LEDCs are mostly in the middle stages of Figure 1.
9. LEDCs are mostly in the last stage of Figure 1.

Investigating migration using a GIS

You need to be able to:
- use GIS to find migration data for an area of in-migration in an MEDC
- use appropriate digital graphing and mapping techniques to present the data
- analyse and interpret the data
- evaluate the GIS technique.

GIS means Geographic Information System. This is a way of storing information about places so that lots of data can be compared and brought together.

The most useful GIS for this is the NINIS website, which is a GIS storing information about different areas of Northern Ireland, which is part of an MEDC experiencing in-migration from Eastern Europe. Go to www.ninis.nisra.gov.uk/mapxtreme/InteractiveMaps.asp. Here you can select information and produce maps.

A GIS is useful because it allows us to access the information quickly and easily, and identify patterns clearly. Lots of information can be linked for one area. This is much quicker than finding the information separately and trying to map it all.

You do not need to memorise the information you find. However, you still need to make sense of the information using what you know about migration.

Impacts of international migration

You need to be able to:
- know about the positive and negative impacts of international migration
- know about migration in one country in the EU, including how many migrate, and the impact on services and the economy.

- **Migration** is when people move house permanently. People who migrate are called migrants.
- **Immigration** is when people move into a country. People who do this are called immigrants.
- **Emigration** is when people move out of a country. People who do this are called emigrants.

Migration can have positive or negative impacts on the country people leave, the country they go to and the migrants themselves.

Get Active

The table on page 65 shows impacts of migration. For each decide whether it is positive or negative, and write a + or – in the last column.

Impact on	Impact	+ or –
The country people leave	• Lose best-qualified people (for example, the brain drain from Northern Ireland to Britain and North America) • Have fewer people to provide for • Emigrants may send money back	
The country people go to	• More people to provide housing, heathcare, education for • More people to work and earn money • Migrants may bring in important skills • Migrants may be willing to work for low wages (for example, food processing plants in Dungannon use migrant workers from Portugal) • Migrants may take jobs away from local people, which may lead to racial tension	
The migrants	• May get good job, be able to send money home • May be lonely, unable to speak the language • May find cultural differences difficult • May be eligible for benefits • May get better healthcare and education than at home	

● Case study: Evaluate impacts of international migration – in the UK

You need to be able to:
- know how many migrate, where from and where to
- evaluate positive and negative impacts on services and the economy.

Numbers and places

You have to learn numbers for this section – but try to stick to one or two easily remembered figures. If you can remember more, that's great.

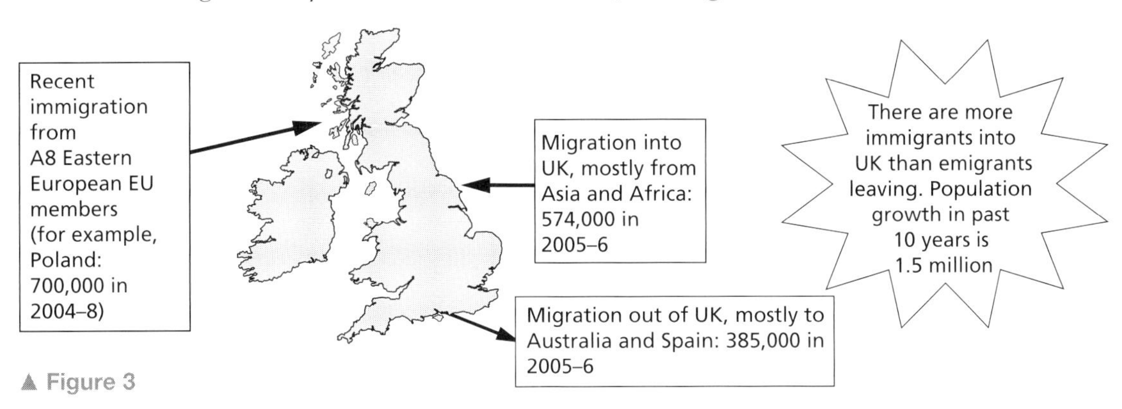

▲ Figure 3

Get Active

The table below shows some of the impacts of international migration on services and the economy in the UK. Evaluate each impact, by deciding whether it is positive or negative for the UK, and labelling + or – in the last column.

Impact on	Impacts	+ or –
Services	• Increase need for housing	
	• Pressure on school places	
	• Language barrier – interpreters needed	
	• Many work in low paid services doing jobs UK-born workers do not want	
	• 1999–2000 £28.8 billion spent on benefits and services for immigrants	
The economy	• Work in factories and increase production	
	• Send home £10 million each day from UK	
	• 1999–2000 £31.2 billion paid to UK government in taxes	
	• Compete with UK-born workers for low-paid jobs	
	• Fill skills gap, doing jobs many UK-born workers will not or cannot do	
	• Push house prices up	

Population structure and pyramids

You need to be able to:

- know about population pyramids including dependency, and the impacts of youth dependency and aged dependency
- refer to two case studies – an MEDC and an LEDC
- interpret different population pyramids.

Population structure, or **composition**, is the way the population is divided between male and female, and different age groups.

It can be shown clearly on a population pyramid such as Figure 5.

Get Active

Match up the following labels to the appropriate points on the population pyramids in Figure 4.

1 Wide base shows high birth rate.
2 Narrow base shows low birth rate.
3 Straight pyramid shows low death rate.
4 Triangular pyramid shows high death rate.
5 Tall pyramid shows high life expectancy.
6 Short pyramid shows low life expectancy.
7 Missing young adult males show out-migration.
8 Extra young adult males show in-migration.

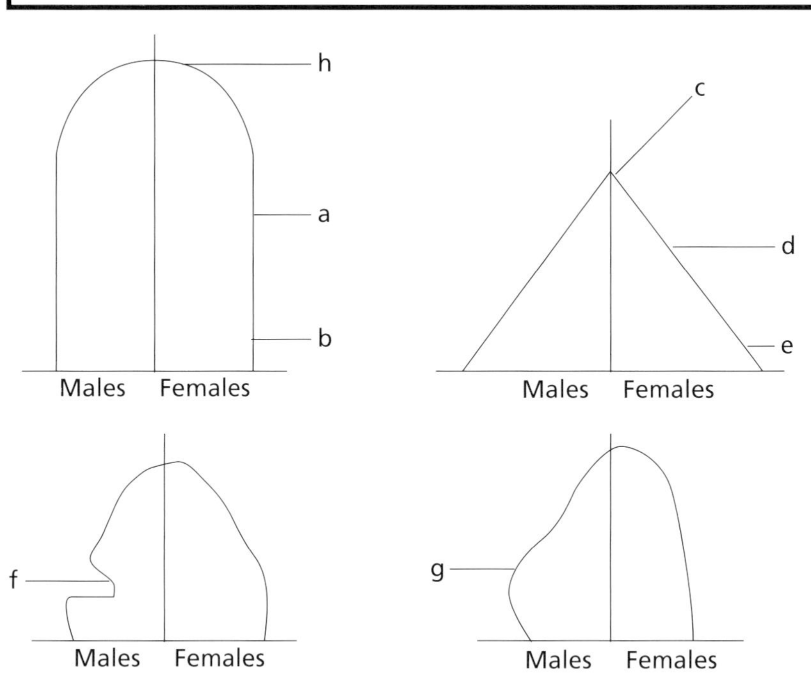

▲ **Figure 4** Population pyramids

Dependency can be seen clearly on population pyramids.

There are two groups of people who are dependent on the people who are of working age to support them:

- aged dependent – age 65+
- youth dependent – age 0–14

We can calculate a dependency ratio to show the percentage of the population dependent on the rest:

$$\text{Dependency ratio} = \frac{\text{Youth-dependent + aged-dependent}}{\text{working population}} \times 100$$

Case study: Population structure of an MEDC – aged-dependent population

Get Active

Match up the following labels to the letters A–C on the pyramid in Figure 5.

1 Tall pyramid shows large percentage of elderly people (16% in 2003). This is the **aged-dependent population**, who do not earn money and need to be supported by the working population.
2 Base of pyramid is getting narrower. This shows the birth rate is falling.
3 Wide section in the middle shows a large population currently of working age who will add to elderly population over the next 30 years.

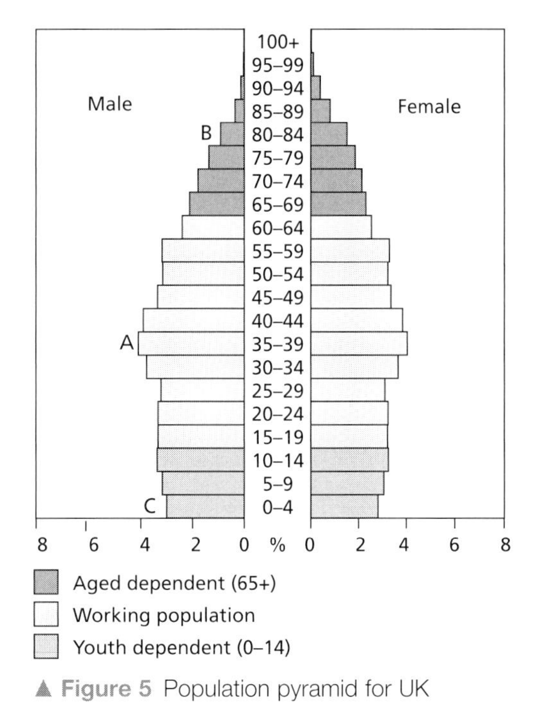

▲ **Figure 5** Population pyramid for UK

Case study: Population structure of an LEDC – youth-dependent population

Get Active

Match up the following labels with the letters A–C on the pyramid in Figure 6.

1 Base of pyramid is getting wider. This shows the birth rate is high. This is the **youth-dependent population**, who do not earn money and need to be supported by the working population.
2 Lots of people of childbearing age means the birth rate will stay high.
3 Pyramid is relatively short, which means death rate is still high.

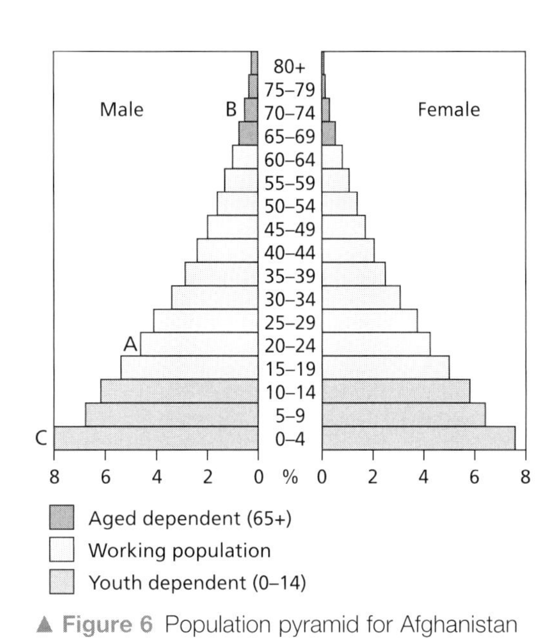

▲ **Figure 6** Population pyramid for Afghanistan

Both aged-dependency and youth-dependency create problems for the country to deal with. The government gets money by taxing people on what they earn. This means people of working age are needed to provide tax money to pay for services like healthcare, education and so on. This creates economic implications (to do with money).

There are also social implications (to do with the way people behave).

Get Active

Make copies of the two tables below and complete them by putting the impacts from the box into the correct part of the tables. Make sure you can explain your decisions. Some may fit into more than one part of the table!

Impacts

1 Adults giving up careers to care for elderly relatives
2 Elderly can provide wise advice
3 Expensive healthcare for the elderly
4 Lack of school buildings and facilities
5 Lack of teachers
6 Large numbers of infant vaccinations needed
7 Lots of young adults entering the labour market
8 Meals on wheels and home helps
9 Pensions
10 Relatives may be able to provide childcare
11 Residential homes needed
12 Strain on carers
13 Strain on primary schools – some operate 2 half-day sessions for different groups of pupils

Socio-economic implications of aged dependency in MEDCs

	Costs	Benefits
Social		
Economic		

Socio-economic implications of youth dependency in LEDCs

	Costs	Benefits
Social		
Economic		

Settlement site, function and hierarchy

You need to be able to:
- know that the original site of a settlement reflected the needs of the people who set it up including defence, water supply and transport
- know the difference between the site and location of a settlement.

Settlement site and location

A **settlement** is a place where people live. Settlements range in size from single farmhouses and small hamlets (groups of houses) to villages, towns and cities.

The **site** of a settlement is the actual spot where it is built, whereas the **location** describes where it is in relation to its surroundings (other settlements, rivers, hills and roads). For example, the site of a settlement might be at a spring for water supply while its location is at the foot of a range of hills, a few kilometres from its nearest neighbouring settlement.

The original settlers needed food, water and shelter. So they selected a site near a stream or spring for drinking water, with land nearby that was suitable for grazing or for growing crops and with supplies of timber for building and for firewood.

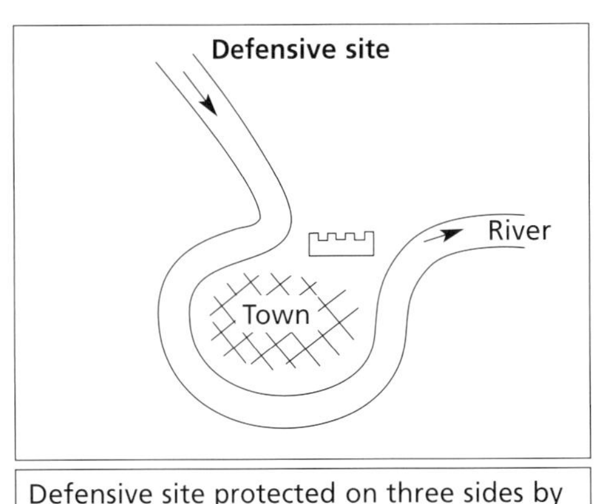

Defensive site protected on three sides by the river and by the castle on the fourth side

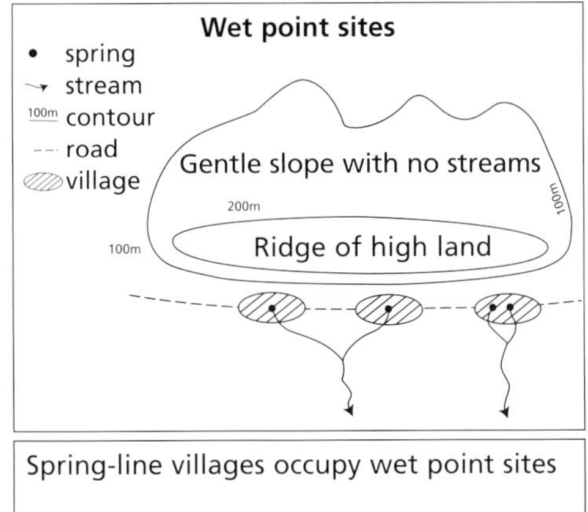

Spring-line villages occupy wet point sites

▲ Figure 7

People who feared attack chose to build on a **defensive site**. They would choose either a site on a hilltop or one surrounded by the meander bend of a river, as these would be easier to defend. A **wet point site** has water supply as its most important factor. In chalk or limestone areas, where most water flows underground, spring-line settlements grew up where springs emerged from the ground.

A **bridging point** is a site where a river could be crossed most easily, and as it became the focus where tracks and roads converged, the settlement that grew up there often became successful as a market town.

Settlement hierarchies

You need to be able to:

- know what is meant by settlement hierarchy
- know that settlements can be arranged in a hierarchy according to their size and function
- know names of places as examples
- define range and threshold.

A **settlement hierarchy** is an arrangement of settlements in order of their size and importance. In a region there may be only one large city, several smaller towns, dozens of villages and hundreds of farmhouses. The numerous small settlements are close together but large settlements, which are few in number, are far apart.

The goods and services found in a settlement can be described as high, middle or low order.

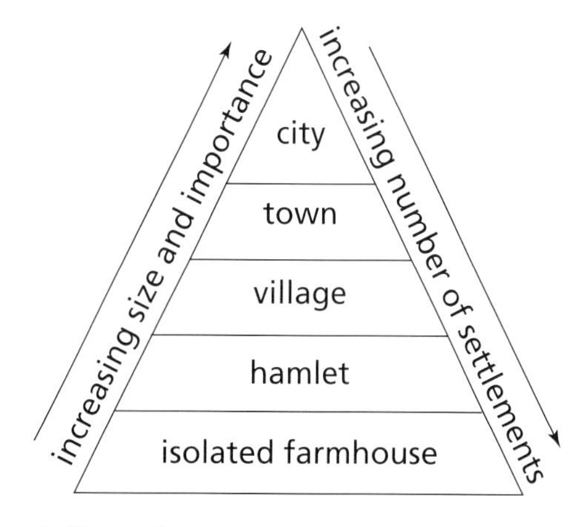

▲ Figure 8

Order of goods or services	Description	Examples
Low order (convenience goods)	• Bought or used frequently, often daily • May be perishable • Found near home	Bread, milk, newspapers
Middle order	• Bought or used less often: weekly, monthly • Willing to travel further to buy or use these	Supermarket shopping, cosmetics, hairdressers, cinema
High order (comparison goods)	• Bought or used seldom, once a year or less often • Durable (long-lasting) • Willing to travel a long distance to obtain these	Christmas shopping, furniture, hospital, theatre

The **sphere of influence** of a settlement is its market area, that is the area from which people travel to use its services. The **range** is the maximum distance that people travel to obtain a service and the **threshold** is the minimum number of people needed to ensure that a service will be able to stay in business. This means that a city like Belfast, at the top of the settlement hierarchy in its region, has a large sphere of influence attracting people from towns and villages all over Northern Ireland to use its wide variety of services. Many of these are high order with a large range and threshold. A village, near the bottom of the settlement hierarchy, has a small sphere of influence as it serves only its own population and the surrounding farmhouses. The services it offers are low order, such as a village shop and primary school, which only need a small threshold population to survive.

The **function** of a settlement is its main economic activity, for example, port, market town, mining town or tourist resort. Large settlements near the top of a hierarchy will have more functions than small settlements near the bottom. People who live in farmhouses and hamlets will need to visit their nearest village to find a shop or primary school, while people from towns and villages will have to go to a city to find a university, museum or major concert venue.

Get Active

1 Match the settlements with their functions.

Settlement	Function
1 Larne – on Co. Antrim coast, a short distance from Scotland.	a) Dormitory town.
2 Ballymena – where eight main roads meet from all over Co. Antrim.	b) Port.
3 Armagh – has museums, a visitor centre and two cathedrals.	c) Tourist town.
4 Hillsborough – close to the M1, making it easy to commute to Belfast.	d) Route centre.
5 Portrush – close to sandy beaches and the Giants Causeway.	e) Cultural centre.

2 Think about the settlement you live in. Is it a city, town, village or smaller settlement? Make a list that includes your settlement and the names of your nearest examples of the other three types of settlement. Arrange your list from the largest to the smallest. An example has been done for you.

City	Town	Village	Smaller settlement
Belfast	Newtownards	Conlig	Craigantlet

Past Paper Exam Questions (Higher Tier)

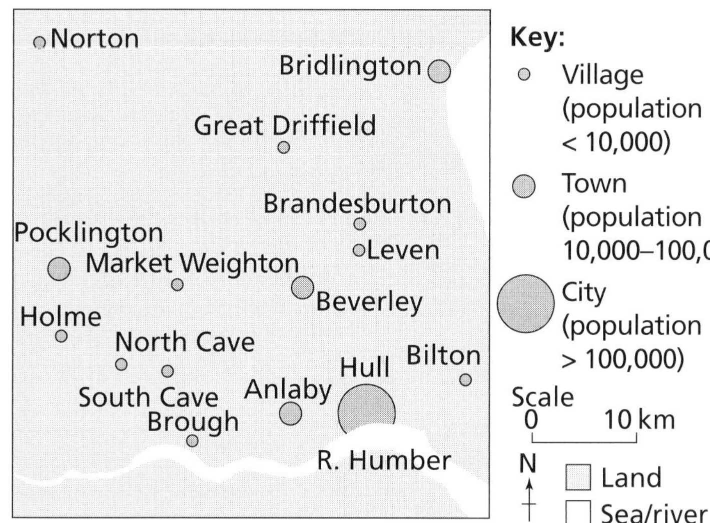

◄ Figure 8

3 (d) Study Figure 8 which shows the settlement pattern in East Yorkshire, England. Answer the questions which follow.

Table 2

Type of settlement	Population size	Total number on map	Example
City		1	Hull
	10,000–100,000		
		10	Holme

(i) Using the information on Figure 8, complete Table 2. [6]

(ii) State **one** possible way in which Hull's location encouraged it to grow into a city. [3]

(iii) State **one** reason why a village has a small sphere of influence. [3]

The map and key provide all the answers. Starting with the city of Hull, the key tells you that Hull's population is over 100,000. On the next line, settlements with 10,000–100,000 people must be towns. You can see four settlements with this symbol on the map and you have the choice of Bridlington, Pocklington, Beverley or Anlaby as an example. Finally, there are ten villages, such as Holme, with fewer than 10,000 people.

This question is designed to find out whether you understand the sphere of influence concept. For 3 marks you might write:

'Villages have few services which are low order such as a pub or general store so they do not attract people from distant settlements to use them.'

Again it is essential to use the map, this time looking for something about Hull's location that has helped it to grow. For 3 marks you need a statement, consequence and elaboration, e.g. 'Hull is located beside the River Humber, so it could develop as a port and profits from trade would help it grow as a city.'

Land use zones in MEDC cities

You need to be able to:

- know what the different zones in cities are
- know where you find each zone in MEDC cities
- know the names of examples of such zones.

Land use means what the land is used for, for example shops, industry or housing. Land use zones are areas which have mainly one land use. The following diagram shows how the main land use zones are commonly arranged in an MEDC city.

1. CBD
2. Inner city; zone of transition and old industrial zones
3. Low-cost, high-density residential suburbs
4. Medium-cost, medium-density residential suburbs
5. High-cost, low-density residential suburbs, modern industrial estates and public housing estates
6. Rural-urban fringe

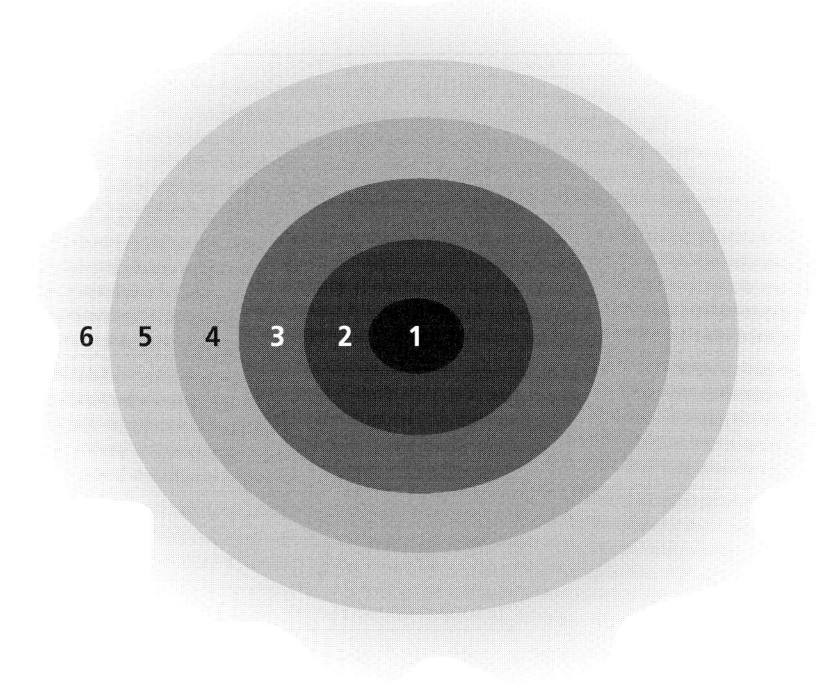

▲ **Figure 9** Land use zones in an MEDC city

Land use zone	Characteristics	Examples in Belfast
Central business district (CBD)	• Shops, offices, entertainment • Easy access by bus, train or car • Land is expensive because space is limited, so people build skyscrapers	Belfast's CBD includes City Hall, Castlecourt and Victoria Square shopping centres, the Grand Opera House and skyscrapers such as the 29-storey Obel building beside the Custom House and M3 bridge
Inner city	• Some buildings old, dilapidated and boarded up, awaiting redevelopment	
	• Some former houses converted into small offices	
	• Large houses divided up, for student housing and so on	• Jerusalem St in the 'Holy Land' near Queen's University
	• Some high-rise blocks of flats	• Divis Flats in West Belfast
Residential suburbs	• Small Victorian terraced houses closest to CBD	• Sydenham (see Figure 21, page 132 in your textbook)
	• Semi-detached houses with gardens built in 1920s and 1930s	
	• Detached houses with gardens and garages, furthest from CBD	• Knock (see Figure 20, page 132 in your textbook)
Industrial zones	• Old industries located close to the docks, railways or canals and to terraced housing for workers	• Neill's flour mill, College Place, BT1 or Harland & Wolff
	• Modern single-storey factories in industrial estates on the edge of the city, on cheaper land and close to motorways for transport	• Mallusk Industrial Estate close to the Sandyknowes junction on the M2
Rural–urban fringe	• A zone of mixed urban and rural land uses	
	• Space available for a golf course, waste recycling centre, hospital or airport	• Roselawn Cemetery, Ballygowan Road, and Ulster Hospital, Dundonald
	• Problems of litter, vandalism or planning blight cause farmland to be under-used	

● Interpretation of maps and aerial photographs

The ideas you have already learned about settlement site, function, land use zones and hierarchy can be applied to a map or aerial photograph of a settlement.

Site	Look for streams as evidence of water supply, contours for a hill-top site or a bridge for a bridging point site
Function	Look for tourist facilities, port facilities or converging roads to indicate that the settlement is a tourist resort, port or route centre
Land use zones	Parallel rows of terraced houses indicate an inner city area; cul-de-sacs of detached or semi-detached houses are found in the suburbs; a town hall is usually in the CBD; large rectangular buildings, perhaps beside a railway or motorway and sometimes labelled *Works* show an industrial zone
Hierarchy	The largest settlement on the map is at the top of the hierarchy and it offers more services, for example, hospital, college, railway station. The many smaller settlements offer lower order services (post office, pub, school)

Get Active

Study the OS map extract on pages 6–7 of *Geography for CCEA GCSE, Second Edition*. What map evidence is there that one function of Cushendall (grid reference 2427) is as a tourist resort?

Urbanisation in LEDCs and MEDCs

● Causes of urbanisation

You need to be able to:
- explain how push and pull factors and natural increase cause urbanisation.

Urbanisation is a process of an increasing percentage of people living in towns and cities. It includes people moving to towns, but this is not the only thing.

a lots of young people migrate to the city, then want to start a family.
b more babies are born alive in towns because of better healthcare.

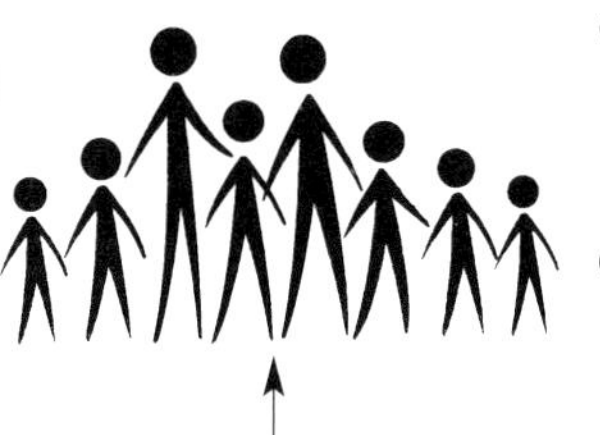

c death rate is lower because there are lots of young people and better healthcare.
d this means natural increase is greater in the city.

Push factors push people away from the countryside

Pull factors pull people towards the cities

▲ **Figure 11** Causes of urbanisation

Get Active

Decide whether each of the following is a push factor or a pull factor:

1 Lack of job opportunities.
2 Good healthcare.
3 Natural disaster.
4 Good education.
5 Good public services like piped water.

● Case study: Shanty town areas – Kolkata, India

People build their own houses because they cannot afford to rent or buy. These areas are called shanty towns. In Kolkata they are called bustees.

You need to be able to:

- describe and explain growth, location and characteristics of shanty town areas
- use place names and figures.

	In general	In Kolkata	Why?
Growth	Slow growth over long period of time, more rapid growth recently	Shanty towns have existed for 150 years. Now growing rapidly, they increased by 32% from 1981 to 1991. Now about four million people live in them	Pull factors – job prospects in city Push factors – mechanisation in farming means people lose jobs in countryside
Location	Cheap land, edge of city, next to main roads, steep slopes	City centre	Old buildings become derelict
		Near factories and main road junctions	Build houses near factories for jobs, or roads for buses
		Vacant land	Cheap unwanted land, less likely to be bulldozed
Characteristics	Poorly constructed, often using scrap wood, or corrugated iron, crowded, few facilities, no street plan, no sewage facilities	Defined as 'unfit for human habitation'. In registered bustees, people have the right to live there. Crowded. Water supply and sanitation shared between many households. Average earnings £7–24 per month – below poverty line	Low wages mean people cannot afford 'normal' houses. Built by occupiers, quickly to meet urgent needs. No planning permission. Electricity and water may only be provided years after building

Get Active

Write out the following list of 'b' words, with at least one piece of evidence for each word. The first one has been done for you. Learn the 'b' words as a chant, and try to remember the evidence that goes with each one.

Bustees are:

- big (four million people in Kolkata, growing at 32% in 10 years)
- badly built
- basic
- beside roads
- bulldozed
- below poverty line.

Urban planning in the inner city

Inner cities are often derelict and in need of rebuilding, so urban planners can design what is built there in order that it works more effectively and is sustainable for the future.

You need to be able to:
- know case study details of one inner city urban planning scheme in an MEDC city
- show how the scheme aims to regenerate and improve the housing, employment and environment of the inner city
- assess how sustainable the scheme is.

Case study: Titanic Quarter, Belfast

Titanic Quarter is a 75-hectare site in Belfast's inner city, within 10 minutes' walk of the CBD, on the east bank of the River Lagan. This is where the Titanic was built in the world-famous Harland & Wolff shipyard but by the 1990s much of the area was disused and derelict. The planned regeneration scheme is estimated to cost £5 billion.

Regeneration means taking action to try to give an area new life – improving the buildings, bringing in new employment and providing social facilities.

Get Active

Copy out the table and place each of the following statements in the appropriate column. Learn the three lists.

Improvements in Titanic Quarter

Housing	Employment	Environment

a 7500 apartments and townhouses to be built.
b Shops, hotels and offices to be built.
c Derelict industrial site will be decontaminated (cleaned up).
d International firms Citigroup and Microsoft have opened offices here.
e Developers claim that 20,000 jobs will be created over 15 years.
f Residents will be attracted by the restaurants, local shops and leisure facilities included in the plans for the site.
g Former slipways where *Titanic* and *Olympic* were built will be laid out as parks and open space for residents and tourists.
h Belfast Metropolitan College and Public Records Office have relocated here.
i 1.5 km of attractive water frontage hosted the Tall Ships in 2009.

Is Titanic Quarter sustainable?

Sustainable development means that the area should change in a way that meets the needs of today but also the needs of future residents. The developers are planning for Titanic Quarter to be sustainable in the following ways:

1 Developing a brownfield site means less countryside has to be built on at the edge of the city. More open space, pleasant views and wildlife can be protected.
2 Providing housing in the inner city means more people live close to their work. This means less fuel is used and there is less pollution.
3 Titanic Quarter is easily accessible and will be served by buses, walkways and cycle paths. A tram route may be introduced in future.
4 Some shipyard buildings will be restored and re-used – the Drawing Offices for public events and the Paint Hall for a film location.
5 Buildings are being designed to use less energy so that less carbon dioxide is released into the atmosphere.
6 Local communities in Belfast's inner city will benefit from the development. The Stepping Stone Project, for example, helps long-term unemployed people from East Belfast to find work.

The scheme will be less sustainable if:

- fewer jobs are actually created than the developers claim, or
- most jobs in Titanic Quarter are taken by well-qualified people who live in the suburbs and commute to work.

Get Active

1 Pick three of the following words or phrases and *explain* how they help to make Titanic Quarter sustainable.

Restoration and reuse	Brownfield site
Accessibility	Energy efficiency
Community development	Inner city housing

2006 Past Paper Exam Questions (Higher Tier)

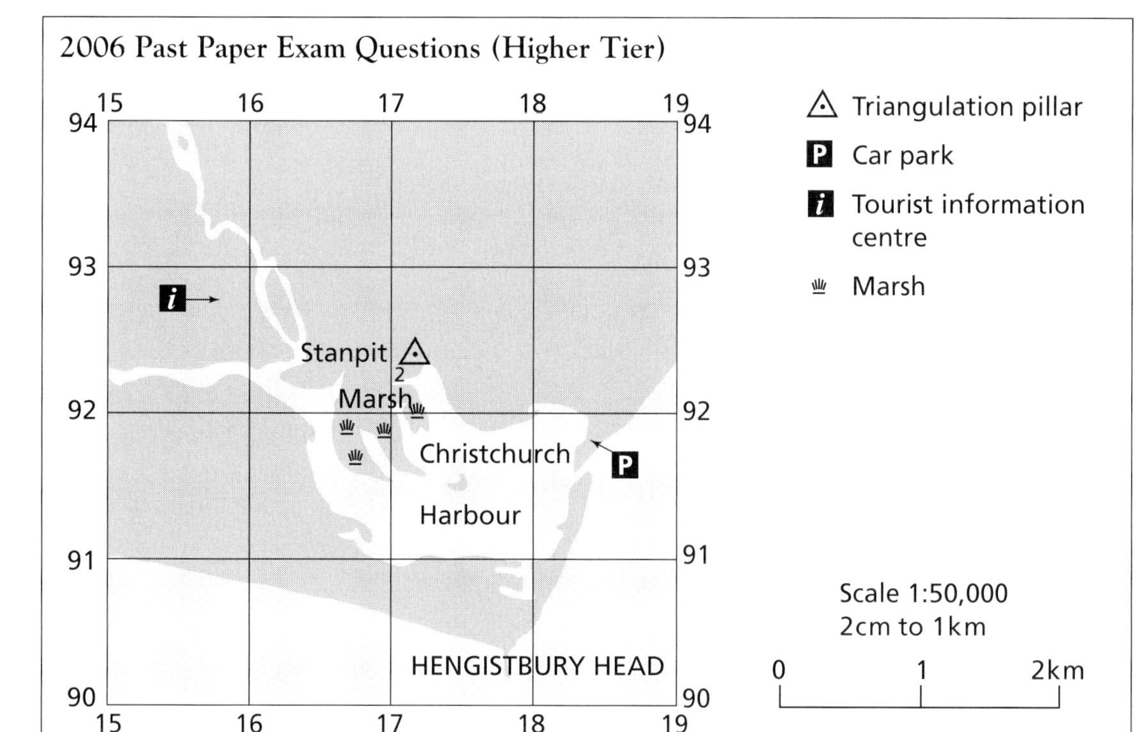

1 (a) Study the Ordnance Survey extract of Christchurch in the South of England where 30% of the population is of retirement age. Answer the questions which follow.

- (i) State the height of the land in Stanpit Marsh, GR172924. [1]
- (ii) State the straight line distance from the tourist information centre at 157928 to the car park at 184918. [2]
- (iii) State the direction of Hengistbury Head, GR1890 from the tourist information centre at 157928. [1]

By using the 6-figure grid reference you should find the triangulation pillar with the height value of 2 metres.

Note the question asks what is the direction of Hengistbury Head from the information centre. Imagine yourself standing at the information centre. What direction do you travel in order to reach Hengistbury Head? The answer is south east. Don't be tempted by more complicated answers such as SSE or ESE. The direction questions only have answers of N, NE, E, SE, S, SW, W or NW.

First, use the grid references to find the car park and information centre. As in both cases there is an arrow from the symbol to the location of the feature it is important to measure the distance in centimetres between the arrowheads, not the symbols. This is 5.8 cm and using the scale you can divide by two to find the distance is 2.9 km. There are 2 marks for an accurate answer of 2.8–3.0 km and 1 mark for the less accurate answers 2.7–2.79 or 3.01–3.1 km.

Knowledge tests

Knowledge test I (Pages 58–65)

1 When birth rate is much greater than death rate, does population grow rapidly, grow slowly or decrease?
2 When birth rate is slightly less than death rate, does population grow rapidly, grow slowly or decrease?
3 What is an emigrant?
4 Suggest one *positive* impact of migration on the economy of the country that people leave.
5 Suggest one *negative* impact of migration on the economy of the country that people leave.
6 What age groups are defined as:
a) youth-dependent and
b) aged-dependent?
7 A wide-based triangular pyramid shows:
a) youth-dependency or
b) aged-dependency?
8 What type of dependent population results in a shortage of school buildings and teachers?
9 What type of dependent population results in increased need for pensions?
10 Suggest one social benefit of having an aged-dependent population.

Knowledge test II (Pages 66–72)

1 What is the term for the actual spot where a settlement is built?
2 Where might a settlement be built to make it easy to defend?
3 Why is a settlement at a bridging point likely to grow?
4 What is meant by settlement hierarchy?
5 True or false? There are more small settlements than large ones.
6 What term means the maximum distance that people travel to obtain a service?
7 What function might a settlement have because it is situated at the coast?
8 What does CBD stand for?
9 Name the land-use zone closest to the CBD where there may be derelict buildings and some high-rise flats.
10 Why is the edge of the city a good location for modern industry?

Knowledge test III (Pages 73–77)

1. Other than push and pull factors, what explains the rapid growth of LEDC cities?
2. If farm workers move to Kolkata because more machinery is being used on farms instead of workers, is this a push factor or a pull factor?
3. If people move to Kolkata because of better job prospects, is this a push factor or a pull factor?
4. What term is used for shanty towns in Kolkata?
5. Approximately how many people live in Kolkata's shanty towns?
6. List all the words you can, beginning with the letter B, to describe Kolkata's shanty towns.
7. Why was Belfast's Titanic Quarter in need of redevelopment?
8. Name two international firms which have opened offices in Titanic Quarter.
9. How many and what type of residences are planned for Titanic Quarter?
10. What features of Titanic Quarter are designed to provide an attractive environment?

Contrasts in World Development

The development gap

● Differences in development between MEDCs and LEDCs

You need to be able to:
- know the differences in development between MEDCs and LEDCs
- know how we measure development using economic and social indicators.

MEDCs are more economically developed countries, or richer countries such as the UK or the USA.

LEDCs are less economically developed countries, or poorer countries such as India or Kenya.

Most MEDCs are in North America and Europe. Most LEDCS are in South America, Africa and Asia. Often a dividing line is drawn on a world map, showing richer countries mostly in the north (but including Australia and New Zealand!) and poorer countries mostly in the south.

▲ **Figure 1** North–south divide

The **development gap** is the difference in development between the MEDCs and the LEDCs. Development is difficult to measure. We use lots of different pieces of information to indicate how developed a country is. These are called development indicators. Some are to do with money – these are **economic indicators**, like how much money is earned. Others are to do with the way people live and their **quality of life** – these are **social indicators**, such as the percentage of people who can read and write.

Get Active

Copy and complete the table below by filling in the second column.

Indicators of development

Indicator	Social or economic	MEDCs	UK (MEDC)	LEDCs	Bangladesh (LEDC)
1 Gross national income (GNI) per person – the value of everything a country produces per person		Usually over $10,000 per year	$33,800	Usually less than $5000 per year. Some countries under $500	$1340
2 Life expectancy – how long people live on average		Over 75 years	79 years	Under 60 years	63 years
3 Number of people per doctor		Not many people per doctor, because there are lots of doctors	610	Lots of people per doctor because there are very few doctors	5556
4 Percentage who can read and write (literacy rate)		High. Many schools and qualified teachers	89.4%	Low. Shortage of schools and qualified teachers	43.1%
5 Percentage of workers employed in farming		Low. Farming is highly mechanised	2.2%	High. Farming is labour intensive	54.0%
6 Percentage of the population living in towns or cities		High. Most people live in towns or cities	89.4%	Low. Most people live in the countryside	20.0%

● Comparing indicators

You need to be able to:

- assess the effectiveness of economic and social indicators
- know the advantages of using the Human Development Index (HDI).

An economic indicator such as GNI per person is good for telling us roughly whether a country is 'rich' or 'poor', but it wrongly assumes that everyone in the country has an equal share in its wealth.

Social indicators are also average figures that hide variations in society. For example, average literacy figures hide the fact that more boys than girls go to school in LEDCs. However, it does give us an idea of the level of education that most people get in the country.

Using only one indicator of development can sometimes disguise the real story about a country's development. For example, if most farmers grow food for their own families, not for sale, a country will have a low GNI per person, so it appears to be poorly developed even though its people may be well fed. This is why the HDI was devised. It combines:

- life expectancy as a measure of *health*
- adult literacy/school enrolment as a measure of *education*
- gross domestic product per person (similar to GNI) as a measure of *wealth*.

The HDI is expressed as a figure between 0 and 1, with countries closest to 1 being more developed. The UK's HDI is 0.946 while the figure for Bangladesh is 0.524.

The advantages of using HDI as an indicator of development include:

- It does not rely solely on wealth as an indicator, but helps us get a picture of a country's *quality of life* by including health and education as well.
- By including education, it looks to the country's potential for future development as well as what has been already achieved in terms of healthcare and wealth.
- It reveals how rich countries use their wealth. For example, an oil-rich country, where little is spent on education for the general population, will have a low HDI despite having a high GNI.

Factors that hinder development in LEDCs

You need to be able to:

- know and understand factors which hinder development in LEDCs, including:
  - historical factors
  - environmental factors
  - dependence on primary activities
  - debt
  - politics.
- illustrate your answers with places.

Get Active

The following are all things which hinder development (make it more difficult) in LEDCs.

1. Many LEDCs suffer hurricanes, earthquakes and floods.
2. Some LEDCs have large debts, so spend their money on repayments instead of hospitals and schools.
3. Ecuador owes $10 billion.
4. Some LEDCs have unstable governments, which change frequently, and have corrupt people working for them.
5. In most LEDCs large numbers of people work in primary activities such as farming or mining. This does not earn much money for the country.
6. The 2004 Boxing Day tsunami destroyed large areas in India.
7. European countries such as the UK and Spain in the past took over large areas of the world as colonies. They imported raw materials from the colonies, and the colonies did not get much money for this.
8. Some diseases are common in hot wet climates where LEDCs are.
9. In the past the UK imported its cotton from India.
10. Zambia gets 98% of its income from exporting copper.

Put each statement into the correct category by making five lists under the following headings:

- Historical factors
- Environmental factors
- Dependence on primary activities
- Debt
- Politics

Strategies to reduce the global development gap

You need to be able to:
- describe one strategy attempting to reduce the global development gap
- identify the organisation, core aims and action taken.

Case study: World Vision

A
- To work alongside people in need to improve their lives.
- To encourage people in rich countries to engage with the need in poor countries.
- To respond quickly to disasters like floods and earthquakes.

B
- Research about poverty, and campaigning to improve things for people in LEDCs.
- Provide relief supplies quickly in disasters.
- Work with poor communities to help them plan for schools, clean water, jobs or whatever is needed.
- Encourage people in MEDCs to sponsor (send money for) individual children to provide them with education and the basics of food, clean water and healthcare.

C
- An international charity which works to raise money and help run development projects in LEDCs.

Get Active

1 Match up the boxes **A**, **B** and **C** above with the following headings
   - The organisation (who is doing this).
   - Core aims (what the organisation is trying to do).
   - Action taken (what they do to achieve the aims).

2 Visit www.worldvision.org.uk for more details about this charity and what it does – or investigate another charity you know of.

If you have a different case study, try drawing a similar set of boxes and filling in information under the same headings.

▲ Villagers and Aid workers sinking a well in Cambodia

Factors contributing to unequal development

● Globalisation and development

You need to be able to:
- know how globalisation helps and hinders development
- know one case study from an LEDC or an NIC.

What is globalisation?

What is globalisation?

The way people, goods, money and ideas move round the world faster and more cheaply than ever before

World trade – brands can be sold everywhere – you can buy Coca-Cola anywhere

Countries sell to each other and are therefore **interdependent** – they depend on each other

Industries have become global – individual companies operate in lots of countries as **multinational corporations (MNCs)**. These are very powerful

Economic decisions or **events in one country affect other countries** quickly

▲ **Figure 2** Globalisation is …

Case study: Globalisation in India

You need to be able to:

- know how globalisation helps and hinders development in one LEDC or one NIC.

How has globalisation affected India?

Since the 1990s entrepreneurs have been encouraged to set up businesses, and lots of MNCs have invested in India. There are lots of workers wanting jobs, who will accept low wages. Large numbers of people speak English, and lots of people who emigrated from India to get work are now returning with new skills.

Get Active

Evaluate the impacts of globalisation in India. Decide whether the following are evidence of helping development, or hindering it. Make two separate lists, and learn them.

1. Half of children under 5 years of age in India are malnourished.
2. Life expectancy has gone up from 59 years in 1990 to 63 in 2004.
3. More people in India now have cars, TVs, washing machines and other consumer goods.
4. More imported goods mean there are fewer jobs in factories for those with little education.
5. MNCs have created many new jobs in call centres and hi-tech industries.
6. Western-style clothes and behaviour are considered shocking by some.
7. Adult literacy rates have increased from 50% in 1990 to 61% in 2004.
8. Increase in the number of enormous shopping centres.
9. Unrest in rural areas has led to guerrilla fighting.

If you have a different case study, make sure you have two separate lists for it, and learn them!

Newly industrialising countries (NICs)

These are countries with rapidly growing economies. They were LEDCs, and are on the way to becoming MEDCs. At first, most of them specialised in electronic goods like TVs and DVD players. You could try remembering BrICK: **Br**azil, **I**ndia, **C**hina, South **K**orea.

● World trade and development

You need to be able to:
- understand how world trade patterns create problems for LEDCs
- refer to a few places to illustrate this.

Trade is the buying and selling of goods and services between one country and another. Globalisation has meant there has been a big increase in the amount of trade around the world. However, there are lots of problems for LEDCs:

- MEDCs control most of the trade and decide how much to pay.
- Many MEDCs are in 'clubs' which trade with each other, and won't let LEDCs sell them their goods. They try to stop LEDCs selling them goods, by charging taxes and putting limits on how much they can sell. For example, the UK is in the European Union.
- MEDCs make expensive goods, which LEDCs have to pay lots of money for.
- LEDCs mostly sell a few primary goods like food or raw materials. They get low prices for these, and are badly affected if the prices go down. For example, Kenya sells mostly tea and coffee.
- Most LEDCs have to buy more than they sell, so they have a trade deficit and have to borrow money.

Get Active

Two countries are described below. One is an LEDC, the other is an MEDC. Get a friend to read you one clue at a time, and see how quickly you can decide which is which. Explain why, using the bullet points above.

Country A
- Wants to sell goods to Germany and is having trouble.
- Has to import all its tractors.
- Had to borrow lots of money recently.
- Sells mostly bananas and coffee.

Country B
- Trades with a lot of other countries.
- Imports cocoa and makes it into chocolate.
- Exports tractors.

Sustainable solutions for unequal development

Appropriate technology

You need to be able to:

- know what is meant by appropriate technology
- know how one sustainable development project in an LEDC aims to help with economic, environmental and social improvements, using place names and figures
- evaluate the success of the project.

Technology is the method or tool which is developed to carry out a task. **Appropriate technology** is technology which is appropriate to the situation. For example, it would be no good giving tractors to farmers in poor countries if they can't get the fuel to run them – it wouldn't be appropriate. Appropriate technology uses the skills, and suits the needs and level of wealth, of local people. It aims to help with economic, environmental and social improvements.

Get Active

Copy the table. Place a tick in the column by the options which are most likely to be appropriate for LEDCs.

	Option a	✓	Option b	✓
1 Fuel	Uses solar power		Uses coal and oil	
2 Materials	Uses wood from local forests		Uses steel from USA	
3 Maintenance	Local people know how to maintain equipment		Needs lots of experts from Europe	
4 Cost	Costs less than a day's wages		Costs three months' wages	
5 Environmental impact	Causes lots of pollution		Causes very little damage to environment	
6 Jobs	Machinery is made in France		Machinery can be made in the village	

Case study: Sustainable development – Kattumarams in India

Fishermen in Tamil Nadu, India, were having problems. The solution was appropriate technology, using technology to help with economic, environmental and social improvements.

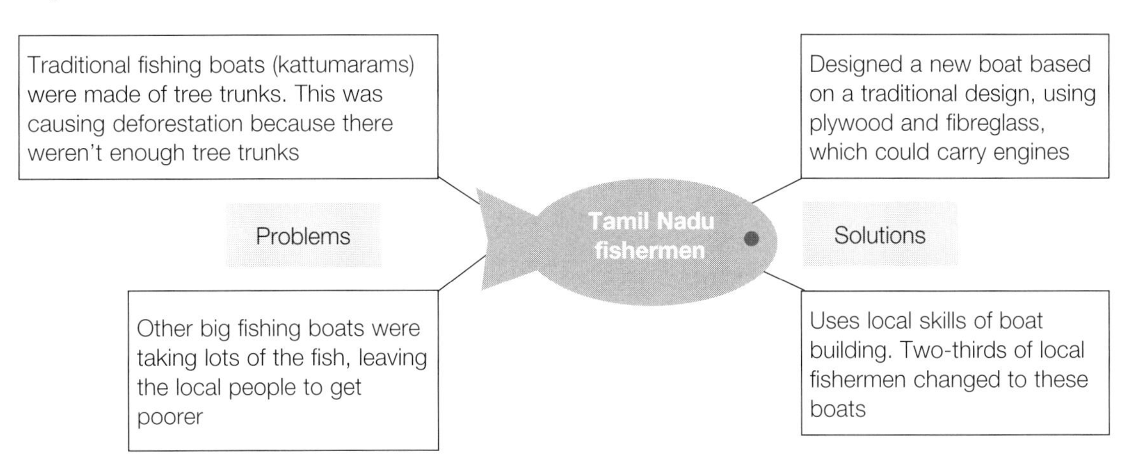

▲ **Figure 3** Problems and appropriate technology solutions

Get Active

1 For each of the following impacts, decide whether it is positive or negative.
2 For each of the positive impacts, decide whether it helps with economic, environmental or social improvements.

Impacts

a Fishermen could continue to catch fish and increase their income	**b** Jobs created in boat building
c Fewer trees are cut down as the new boat is made of fibreglass	**d** Using engines contributes to global warming
e Can bring more fish to the shore than they could in the old boats	**f** As fishermen make more money, they can afford more healthcare, better food, and can send more of their children to school
g Some materials have to be bought from elsewhere, for example, fibreglass	**h** Petrol is scarce and will not last forever

3 Try writing a postcard as if you are one of the fishermen, to your cousin living in Northern Ireland, explaining how your life has changed.

Fair trade

You need to be able to:

- know what fair trade is
- understand the advantages it has for LEDCs, and refer to some places for illustration.

Fair trade means people who make or grow something are paid a fair price for their work. This price is guaranteed, so the producer will not lose out if world prices fall.

If you buy a bar of chocolate which is *not* fair trade, the profits are shared as shown in Figure 4. Fair trade means the producers get more of the money.

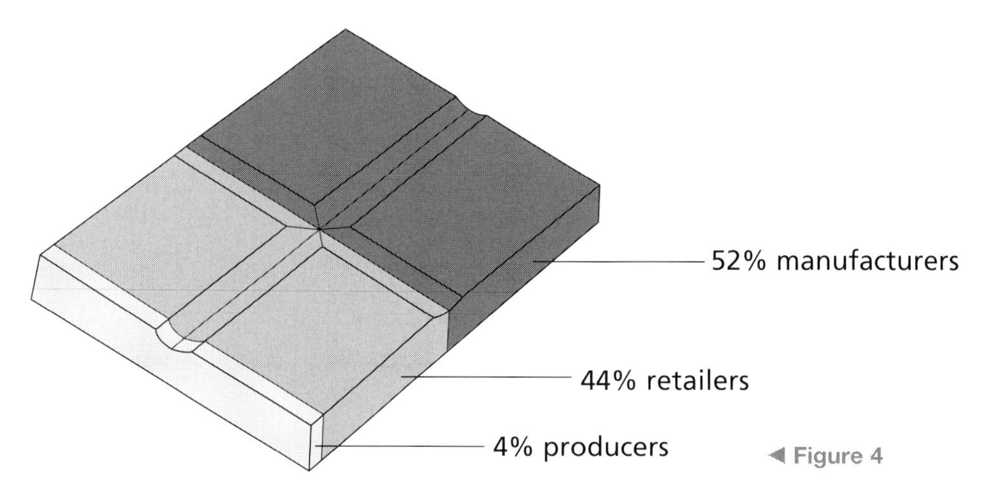

◀ Figure 4

The advantages of fair trade for LEDCs:

- Guarantees a minimum wage for farmers.
- Farmers can provide for their families.
- Farmers have access to cheap loans.
- Farmers control the business.
- Profits are used by groups of farmers to help provide healthcare, education and transport for their communities.
- Encourages sustainable farming practices.

The island of St Lucia in the West Indies has 13 groups of fair trade banana farmers. Since 2000 they have used part of their income from selling fair trade bananas to build a community centre, provide science equipment in two schools and buy a new truck to deliver fertilisers and packaging materials to members.

Get Active

Draw a coffee jar or bar of chocolate (or get hold of a real one!) and label it with:

- bad things about normal trade in black
- good things about fair trade in green.

2006 Past Paper Exam Questions

Higher

2 (d) Study Figure 5a which shows changes in the market price of coffee. Answer the questions which follow.

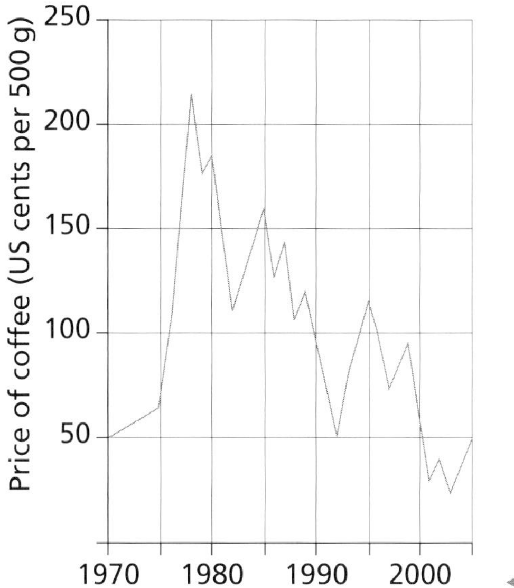

◄ Figure 5a

(i) Describe the changes in the price of coffee on the world market shown in Figure 5a. [3]

For full marks you need to refer to the price rising until 1978 and then falling unevenly, as well as quoting figures from the graph, e.g. 'Between 1970 and 1978 the price of coffee rose sharply from 50 cents to 215 cents per 500 g. After 1978 the price fluctuated but fell overall, returning to 50 cents in 2005.'

Foundation

(e) Study Figure 5b which shows the price paid for coffee on the world market and the price paid by Café Direct (a fair trade company) in 2005. Answer the questions which follow.

	World market price (US cents per 500 g of coffee)	**Café Direct pays** (US cents per 500 g of coffee)
Arabica coffee	120 cents	132 cents

▲ Figure 5b

(i) Explain how Café Direct (a fair trade company) helps coffee growers. [3]

This question refers to a table so you should use information from it as well as focusing on how the higher fair trade price helps the growers. 'The growers get an extra 12 cents for every 500 g of coffee they sell to Café Direct, so they have more income to spend on educating their children.'

(ii) State **one** reason why people in MEDCs would want to buy Café Direct coffee rather than cheaper brands of coffee. [3]

Think about **one** reason but add elaboration and consequence to gain 3 marks. 'Some people in MEDCs want to buy Café Direct coffee because it makes them feel better to know that coffee growers are being paid fairly and so are able to get better education and healthcare.'

Aid

You need to be able to:

- know about different types of aid
- know how aid can bring both benefits and problems to LEDCs, and refer to places for illustration.

Aid is resources given by one country or organisation to another country. These resources can include:

- money (given or loaned)
- expertise (people such as engineers, doctors, teachers)
- goods (food, technology, equipment such as tents or blankets).

Types of aid

- Short-term aid is given when there is an emergency such as a flood.
- Long-term aid is given to help a country develop.
- Bilateral aid is when one country gives aid directly to another.
- Multilateral aid is when lots of governments give money to world organisations such as the UN, who then give it to countries that need it.
- Tied aid is when the country giving the money tells a country what to spend it on.
- Voluntary organisations are charities such as Comic Relief or Oxfam. They get their money from people rather than governments.

Get Active

Decide what type of aid is happening in each of the following examples. Some may be more than one type.

1. Many people contributed to a fund to help people in Sri Lanka after the 2004 Boxing Day tsunami (two types).
2. Ireland gives money to Mozambique (one type).
3. Wealthy countries pay money into the World Bank, and it gives loans to Tanzania and Pakistan for irrigation projects (two types).
4. Granny is sponsored to abseil down the Europa Hotel in a chicken costume for Comic Relief, who send the money to Kenya to help women plant trees for fruit, firewood and building materials (two types).

Get Active

Below are possible outcomes of aid. Decide whether each is a benefit or a problem.

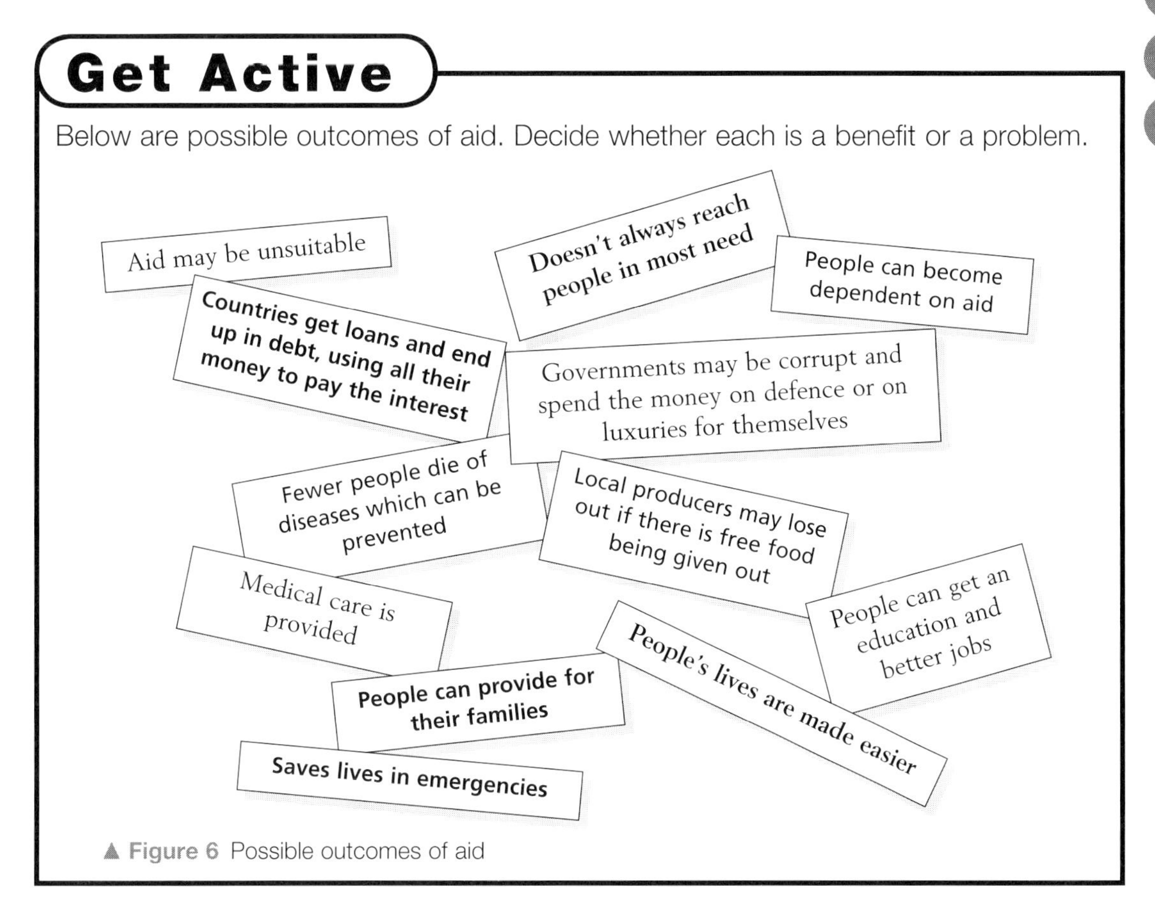

▲ **Figure 6** Possible outcomes of aid

Knowledge tests

Knowledge test I (Pages 80–84)

1 Are the following measures of development economic or social:
   a) literacy rate,
   b) life expectancy,
   c) gross national income?
2 What type of indicators, social or economic, aim to measure quality of life?
3 Does a small percentage of the workforce employed in farming indicate an LEDC or an MEDC?
4 Does a small number of people per doctor indicate an LEDC or an MEDC?
5 Name the development indicator that combines measures of health, education and wealth.
6 What does it use as measures of:
   a) health,
   b) education and
   c) wealth?

7 Which countries, LEDCs or MEDCs, generally export primary products such as food crops?
8 True or false? One reason for LEDCs having developed only slowly is that in the past they were colonies.
9 True or false? One reason for LEDCs having developed only slowly is that they often suffer from hurricanes, earthquakes, floods and droughts.
10 State two actions that World Vision takes to reduce the global development gap.

Knowledge test II (Pages 85–87)

1 What is meant by the term globalisation?
2 MNCs are one aspect of globalisation. What are MNCs?
3 How has globalisation benefited India?
4 What problems has globalisation caused for India?
5 What do the letters NIC stand for?
6 True or false? NICs are the same as LEDCs.
7 What term means buying and selling goods and services between one country and another?
8 Which countries, LEDCs or MEDCs, mostly import food and raw materials?
9 Which countries, LEDCs or MEDCs, mostly import manufactured goods?
10 Which countries, LEDCs or MEDCs, are badly affected if world prices of primary products (tea, coffee, cotton) fall?

Knowledge test III (Pages 88–93)

1 What is appropriate technology?
2 State two *economic* improvements that appropriate technology brought for fishermen in Tamil Nadu, India.
3 State one *environmental* improvement resulting from the Tamil Nadu project.
4 State one *social* improvement resulting from the Tamil Nadu project.
5 State one negative impact of the Tamil Nadu project.
6 How might a farmer's family benefit from fair trade?
7 What fair trade products are easily available to buy in supermarkets?
8 What type of aid is given by many people to help those suffering as a result of an earthquake or famine?
9 What is the purpose of long-term aid?
10 What is meant by bilateral aid?

Managing Our Resources

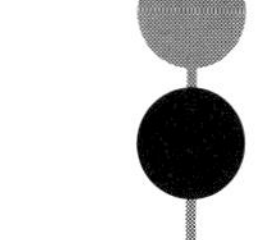

The impact of our increasing use of resources on the environment

● What are resources?

A **resource** is something that we use, such as water, farmland, fuel, housing. Resources such as coal, oil and natural gas are **finite** or **non-renewable** as they will run out one day. **Renewable** resources are either continuous (like **wind** and **solar power**) or can renew themselves naturally if not abused (like trees). If a country earns money from tourism then an unspoiled coastline or a rich culture are also important resources.

● Carbon footprints

You need to be able to:

- know what is meant by carbon footprints
- know how our use of resources has impacts on the environment.

Using *any* resource has an impact on the environment. If we eat cornflakes for breakfast we have to consider the power used to plough the soil and harvest the cereal crop, chemicals used in pest control, power used in the cornflakes factory, energy used in growing and felling trees and producing the cardboard packaging, transport to the factory and then to the supermarket and our homes. Each of these stages has an impact on the environment.

The fuel burned in vehicles and power stations produces carbon dioxide, contributing to greenhouse gases in the atmosphere as well as the **depletion** of our reserves of fossil fuels. The chemicals used in farming may pollute the soil and nearby streams and potentially harm wildlife.

The **carbon footprint** is a measurement of all the greenhouse gases that we each produce, converted into the equivalent weight of carbon dioxide produced so that they can be compared. Research has shown that richer people tend to have larger carbon footprints – people with more money spend more time in bigger cars, eating more exotic food (imported from around the world), holidaying in distant countries and buying more consumer goods which all contribute towards a larger carbon footprint.

Get Active

Caroline Careless has a carbon footprint of 12.56 tonnes but Gregory Green's footprint is only 2.00 tonnes. The table lists some of the choices they each have made about how they live. Study the list and decide who made each of the choices listed – tick either Caroline or Gregory.

Lifestyle choice	Caroline	Gregory
a House has no insulation in loft or walls		
b Only low-energy light bulbs are used		
c Water heated by solar panels		
d TVs and computers always left on stand-by		
e Hangs out washing to dry on clothes line		
f Washes clothes at 30°C		
g Car used for short trips to local shops		
h Rarely uses public transport		
i Walks or cycles to local shops		
j Two long-haul holiday flights per year		

Specimen Exam Questions (Higher Tier)

3 (a) (i) State the meaning of the term **carbon footprint**. [2]

(ii) For a named MEDC city, describe **and** explain how **one** traffic control measure contributes to the sustainable development of the city. [6]

To get two marks for this answer, you need to mention greenhouse gases, for example, 'the total greenhouse gases caused by a person, converted into the equivalent weight of carbon dioxide.'

You need to pick one traffic control measure to describe, and then explain how it helps make Freiburg more sustainable. Two facts or figures must be included for top marks.

Case study: Measures to manage traffic in a sustainable manner – Freiburg, Germany

You need to be able to:

- know what has been done to manage traffic in one city in the EU (not the British Isles)
- evaluate how successful these measures have been in terms of sustainability.

Freiburg – solutions to traffic problems

The city – 200,000 people, Southern Germany, historic city centre within old walls

Problems – population growth leading to increased traffic congestion in narrow city centre streets

Solutions

- Pedestrianised city centre
- Improved public transport, very cheap fares on non-polluting trams
- Park and ride facilities on edge of city for commuters
- 200 miles of cycle paths, cycle storage at tram stops so people can 'bike and ride'
- No free car parking in city centre
- City is compact, 70% of people live near a tram stop

Impacts

- 4000 fewer cars each day than in 1970
- 70% of local trips made by tram
- Public transport use more than doubled since 1980
- All this reduces congestion, fuel use and pollution, ensuring that Freiburg's transport system is becoming more sustainable
- People travelling in the wider region are still affected by policies that encourage car use
- Even with all these policies people still use cars in Freiburg

Increasing demand for resources

● Demand for resources in LEDCs

You need to be able to:

- understand how population growth and economic development in LEDCs increase demand for resources
- understand how this puts pressure on people and the environment
- know how this works for one case study in an LEDC.

Case study: China

Population growth

China's population is growing at one million per month in spite of the one-child policy.

There are around 1.3 billion people in China.

Economic development

China has developed factories and sells its goods all over the world.

The economy is growing at 9.5% each year. People have more money to spend. In 2006 average earnings were over five times as much as in 1981.

Increased demand for resources

More people, all wanting more things – washing machines, fridges, cars, food, clothes, TVs. These all have to be made – using metal, energy, cotton and other resources.

Pressure on people

- Air pollution responsible for 750,000 deaths per year – some people wore masks for the Beijing Olympics.
- Traffic congestion – commuting to work in big cities like Beijing often takes 3 hours.
- Water supplies polluted in 90% of cities.

Pressure on the environment

- Serious air pollution in two-thirds of cities studied.
- Waste – four million fridges thrown away each year which releases CFCs, a greenhouse gas.
- Deforestation – 10 million hectares lost in 5 years, contributing to global warming.
- Tigers dying out.

Get Active

Draw six cartoons – a baby, factory, person in a mask, traffic jam, trees and tiger. Label your drawings with two or three key pieces of information. Use the cartoons to help you remember and revise this case study.

Benefits and problems of a renewable energy source

You need to be able to:
- evaluate benefits and problems of one renewable energy source as a sustainable solution to increasing demand for resources
- relate this to one case study of an MEDC.

What is a renewable energy source?

Coal, oil and natural gas are non-renewable (or finite) energy resources. The more of them that are extracted from the Earth and used, the fewer energy reserves there are left for use by future generations. This is not **sustainable**. Burning coal, oil and gas also contributes to climate change and acid rain so that the environment is damaged for the future too. Using renewable energy resources, which will not run out, is therefore more **sustainable**. There is a UK government target of generating 15% of all electricity by renewable methods by 2015.

Some renewable energy sources are:

- **Wind power** – where the wind turns a turbine, which generates electricity.
- **Solar power** – where the sun's energy causes chemical reactions which generate electricity.
- **Biofuels** – where plants such as willow trees, maize or sugar are burnt or processed to create energy.

Case study: Wind power in Denmark

Wind power is very important in Denmark:

- It produces 19% of the country's electricity.
- The Danes plan to increase wind power to produce 50% of their electricity by 2025.
- Most wind farms are in the sea, about 14 km from the shore.
- The island of Samsø has 11 **onshore** wind turbines (on land) and 10 offshore (in the sea).
- Horns Rev, the largest wind farm, can generate enough electricity for 150,000 private homes, or nearly 2% of Denmark's total electricity consumption.

Get Active

The following table has a list of benefits and problems associated with wind power in Denmark. Evaluate each with + for benefit and – for problem. Make separate lists of the benefits and problems, and learn them!

	+ or –
a Wind farms in the sea will damage the plants and animals on the sea bed	
b No greenhouse gases to contribute to global warming	
c Wind speed will vary, but it will never run out	
d Wind turbines can be seen up to 45 km away, and some people dislike them	
e Some birds seem to avoid areas with turbines, so may lose their feeding grounds	
f When wind speed drops, no electricity is produced, and it is difficult to store electricity	
g Electric cars could be charged at night when there is spare wind power, and could store the electricity	
h Tourists may stop visiting places near wind farms	
i Denmark will have to import less fuel	

Managing waste to protect our environment

You need to be able to:
- understand what waste management is
- understand why it is a major issue in the UK.

Waste management is a major issue

Waste management is how litter and other waste is dealt with. For a long time, waste in Northern Ireland has mostly gone to **landfill sites**, where it is simply buried underground in old quarries or natural dips in the ground.

There are three reasons why this is a problem:

- Shortage of landfill sites – lots are full already, and people don't like new ones near where they live.
- Environmental and health concerns – people worry that chemicals from the waste will poison groundwater and spread disease.
- The need to meet government targets – the EU will fine countries which don't reach them! By 2020 the UK aims to have less than a third of its waste going to landfill.

The waste hierarchy

You need to be able to:

- describe the waste hierarchy
- describe the idea of the 3Rs – reduce, reuse, recycle
- know about a case study of waste management.

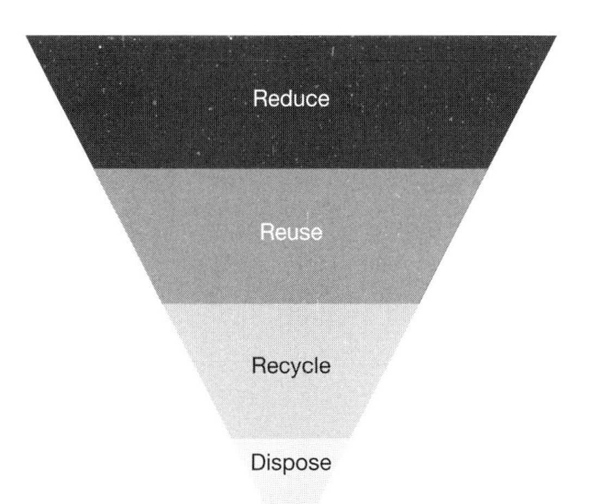

There are lots of ways of dealing with waste which are more sustainable than landfill. The **waste hierarchy** lists these from most sustainable at the top, to least sustainable at the bottom.

◀ **Figure 1** The waste hierarchy

Get Active

1 Match the following waste management options with the descriptions below them.

Composting   Incineration (waste to energy)   Landfill   Recycling   Reducing   Reusing

Descriptions

a Putting all types of rubbish in a hole in the ground.
b Burning waste and using the heat as energy.
c Turning a used product into something new, like old paper into cardboard.
d Using something again, such as a 'bag for life'.
e Buying less and therefore wasting less.
f Allowing garden and food waste to rot and make compost.

2 Put the options in order, from most sustainable to least sustainable.

There have been campaigns to encourage people to use the most sustainable of these options – reduce, reuse, recycle.

Get Active

For each of the following situations, decide whether the people are reducing, reusing or recycling. Some may be doing more than one!

- **a** Gill's dad always keeps paper that has been printed on one side to use as scrap paper.
- **b** Jonathan has bought rechargeable batteries for his camera.
- **c** Bob the Builder always takes broken metal from old buildings to the scrapyard where it will be melted and turned into new metal.
- **d** Oliver always takes his lunch to school in a lunch box so he doesn't need clingfilm.
- **e** Theo puts all the empty cans in the blue bin instead of the grey one.
- **f** Mum bought low-energy light bulbs for the whole house. She said they should each last 6 years.
- **g** Deborah's new pencil case was made from an old car tyre.
- **h** Susan's baby wears washable nappies instead of disposable ones.

● Case study: Belfast City Council

Belfast produces a large amount of waste – about 121,000 tonnes of household waste per year. Until recently most of this has gone to landfill. Now there is a target to reduce landfill by 80% by 2020.

In 2006–7 24% was recycled or composted. This is increasing. Houses in Belfast now have blue bins or black boxes for paper, card, cans and plastic bottles. These are collected for recycling. Most also have brown bins for grass cuttings and other garden waste, which are taken for composting. Subsidised home compost bins are also available.

There are still 'superdumps', or landfill sites such as one at Mallusk. This is unsustainable, so the Eastern Region Waste Management Group have plans to build a waste-to-energy plant (incinerator), which would heat 20,000 homes, and a mechanical biological treatment facility, which will help sort out material and compost where possible.

Get Active

Every time you put something in the bin today, make sure you put it in the correct bin, and remind yourself where it will be going and how sustainable it is!

Sustainable tourism to preserve the environment

● Tourism growth

You need to be able to:
- explain the reasons why tourism has grown globally since the 1960s.

In the 1960s a typical family would feel fortunate to spend a week on holiday at a seaside town in this country. Since then, there has been an increase in the number of holidays taken per year and the distance travelled. There are a few reasons for this:

- *Increased leisure time.* Workers now get more days' annual holiday so may take short-break holidays as well as the usual 'summer holiday'. People who have flexible working can arrange to take a long weekend break.
- *Increased disposable income.* The incomes of workers have grown faster than the cost of living. Most people now have a greater 'disposable income': the money they can choose to spend, once they have paid for the essentials. This means that many people can afford to travel much further afield than in the past.
- *Cheaper travel.* Low-cost airlines such as easyJet and Ryanair have dramatically reduced the cost of flying. This makes weekend breaks in foreign cities more affordable, and has encouraged some people to buy second homes in Mediterranean countries, travelling cheaply and conveniently several times per year. Cheap flights also encourage travellers to plan longer journeys that would once have seemed too expensive.
- *Increased health and wealth of pensioners.*

Date	Population aged over 60 years (%)
1920	10
1940	16
1980	18
2005	21

Older people are able to travel more than in previous generations as they are living longer and in better health. They are not restricted to the popular school holiday times so they can also take advantage of cheaper off-peak holiday prices. If they have saved for their retirement they can sometimes afford several holidays a year.

Impacts of tourism

You need to be able to:

- evaluate positive and negative economic impacts of tourism
- evaluate positive and negative environmental impacts
- evaluate positive and negative cultural impacts
- refer to places when you write about this topic.

Impacts	Positive	Negative
Economic	Jobs created in hotels, restaurants, taxis, souvenir shops. People employed in tourism spend their income in the local shops and services, boosting the local economy.	Money earned from tourism is often lost to the local economy because, for example, hotel profits are sent to the owners overseas. If tourism uses migrant workers, they may send some of their wages home so local spending is reduced.
Environmental	New airports and better roads aimed to benefit tourists actually improve the environment for everyone. Local councils often take better care of historic buildings and scenic countryside when they know these attract tourists.	Water is used in vast quantities for swimming pools and golf courses, and may leave a shortage for farmers and local people. Wildlife and ecosystems can be endangered by the litter, sewage and other pollution created by tourists.
Cultural	Both visitors and local people get the chance to learn from each other's culture.	Local young people can become involved with drunkenness and crime associated with some tourist resorts. Traditional costumes and dancing become detached from their genuine culture and devalued when performed to earn money from visitors.

Get Active

Read the following list of tourism impacts carefully. For each one, decide whether it is positive or negative and whether it is economic, environmental or cultural.

Impacts of tourism	Positive or negative	Economic, environmental or cultural
1 In Cyprus, when turtles hatch at night from eggs laid on beaches they should creep towards the moonlit sea but are endangered by bright hotel lights which lure them towards roads.		
2 Local residents in Zanzibar use less than 50 litres of water per day but tourists in hotels use 931 litres each per day on average.		
3 Over 200 million people employed in tourism worldwide.		
4 38% of visitors to Ayers Rock climb to the top even though the rock is sacred to the Aboriginal people of central Australia who request visitors not to make the climb.		
5 An estimated 17% of Kenya's tourist earnings 'leaks away' from the Kenyan economy.		

The impact of sustainable tourism on the local community and on the environment

You need to be able to:

- know case study details about one sustainable tourism project in either an LEDC or an MEDC
- assess its impact on the local community
- assess its impact on the environment.

Case study: The impact of sustainable tourism in Nam Ha, Laos

For tourism to be sustainable, it must not damage the environment on which future generations will depend. One way to do this is through **ecotourism** (also known as **green tourism**) where visitors enjoy nature at first hand while protecting the environment and local way of life.

Get Active

Make a large copy of the table below and choose the appropriate statements from the following list to fill each box. Some statements may belong in more than one box. If your case study is different, fill out the boxes with facts about it, and learn them.

a Laos – an LEDC in South-east Asia (population 6.8 million).
b Ecotourism project organised by UNESCO.
c Wilderness area of mountains and deciduous forests.
d Guides and trekkers help to deter poachers so rare species are conserved.
e Nam Ha National Protected Area in north of Laos.
f Boating on Nam Ha River and trekking through forests.
g All trekking and boat trips must use Nam Ha Eco-guide Service.
h Wildlife here includes rare clouded leopards and tigers in danger of extinction, gibbons, Asian elephants and 288 bird species.
i Two proposed roads, which would have led to logging and illegal trade in wildlife, have not been allowed to go ahead.
j Local people become guides instead of hunters so wildlife is conserved.
k Earnings of Nam Ha Eco-guide Service invested in small-scale development to benefit local people.
l Village income increased

1 Facts about the location:	**2** Description of the sustainable tourism project:
• • • •	• • •
3 Impact on local community:	**4** Impact on environment:
• • • •	• • • •

Knowledge tests

Knowledge test I (Pages 95–97)

1 Give two examples of resources.
2 Coal, oil and natural gas are all what type of resource?
3 Give two examples of continuous renewable resources.
4 Explain what is meant by resource depletion.
5 What term do we use for a way of measuring greenhouse emissions created by our activities?
6 Give two ways in which people can reduce their greenhouse gas emissions.
7 Name the main traffic problem which was faced by the city of Freiburg in Germany.
8 Give two measures taken to manage traffic in Freiburg.
9 Give two positive outcomes of these measures.
10 Give one negative outcome of these measures.

Knowledge test II (Pages 98–100)

1 How fast is China's population growing?
2 How fast is China's economy growing?
3 Give one example of pressure on the environment created by increased demand for resources.
4 Give one example of pressure on people created by increased demand for resources.
5 The UK government aims to generate 15% of electricity by what kind of methods by 2015?
6 What is meant by biofuels?
7 How much of Denmark's electricity is produced by wind power?
8 Give two examples of problems associated with wind power in Denmark.
9 Give two examples of benefits associated with wind power in Denmark.
10 True or false? Burning oil is not sustainable.

Knowledge test III (Pages 100–106)

1 Give one reason why landfill is not a good way to manage waste.
2 Which of the following options is the most sustainable:
   a) landfill,
   b) recycling,
   c) reusing?
3 How much household waste does Belfast produce each year?
4 What is a waste-to-energy plant?
5 How can grass cuttings be disposed of sustainably?
6 State two reasons why tourism has grown globally since the 1960s.
7 How can tourism have a negative economic impact?
8 What is another name for "green tourism"?
9 Name three wildlife species which attract tourists to the Nam Ha region of Laos.
10 How do visitors travel in Nam Ha to see the wildlife? (Two methods.)

Answers to Knowledge tests

Unit 1: Understanding Our Natural World

Theme A: The Dynamic Landscape

Knowledge test I (page 25)

1 Precipitation (rain, snow, sleet or hail).
2 Interception.
3 Groundwater flow.
4 Water sinking down through rock.
5 Output.
6 Evaporation.
7 Infiltration sinks *into* soil; throughflow flows *through* soil.
8 Tributary.
9 Confluence.
10 Boundary of a drainage basin or highest point all around a drainage basin.

Knowledge test II (page 25)

1 They get smaller and more rounded as they knock against each other and the banks.
2 Abrasion/corrasion.
3 Saltation.
4 Deposition.
5 Plungepool.
6 True.
7 Sediment/alluvium/river's load.
8 Destructive waves.
9 Longshore drift.
10 Spit formed by deposition rather than erosion.

Knowledge test III (page 25)

1 Heavy rain over North York Moors *or* ground already saturated by previous rain.
2 Cutting peat from North York Moors *or* building houses on the floodplain at Malton.
3 Water for farming; makes soil more fertile; habitat for fish.
4 Afforestation; safe flooding zones.
5 Levees; dams; straightening meanders.
6 Rise in sea level.
7 a) With steps, b) curved.
8 Beach nourishment.
9 Groynes.
10 Gabions and groynes.

Theme B: Our Changing Weather and Climate

Knowledge test I (page 41)

1 False.
2 Anemometer.
3 Rain gauge.
4 Millibars (mb).
5 °C.
6 Cloud cover.
7 Stevenson screen.
8 In the shade away from direct sunlight; vents allow the air to pass through.
9 In the open away from buildings; partly sunk in the ground.
10 Three of land-based stations, balloons, buoys, weather ships, satellites.

Knowledge test II (page 41)

1 Polar maritime, cold and wet.
2 a) High, b) sinking.
3 b) Warm and cold fronts.
4 Depression.
5 Anticyclone.
6 Depression.
7 Cold front.
8 Increased sales of ice cream and suntan lotion in summer; winter fog causes delays to commuters and air travellers so less work is done.
9 Fronts bring rain so crops have sufficient moisture.
10 Flooding caused by heavy rain; damage to buildings/road blocked as result of strong winds.

Knowledge test III (page 42)

1 False.
2 False – volcanic activity and natural climatic cycles also contribute.
3 Carbon dioxide (CO_2).
4 Ice caps melting; sea water expanding.
5 Increased tourism benefiting the economy; increased yields of crops such as vines and maize.
6 a) Kyoto, b) Copenhagen.
7 Two of wind, solar, biofuels, tidal, wave.
8 A charge (e.g. of £8 per visit) to drive into central London, intended to discourage the use of private cars and encourage public transport.
9 Sustainable development of rainforest resources (fruits, nuts, oils and medicines).
10 Trees absorb the greenhouse gas CO_2 as they grow but deforestation increases both CO_2 and global warming.

Theme C: The Restless Earth

Knowledge test I (page 56)

1 Sedimentary.
2 Granite.
3 Limestone, sandstone.
4 a) Slate, b) marble.
5 Convection currents in the mantle.
6 Constructive.
7 Destructive margin.
8 Constructive margin.
9 Subduction.
10 Plates slide past each other.

Knowledge test II (page 56)

1 Co. Antrim/Antrim Plateau.
2 False – repeated eruptions from fissures or cracks.
3 False – slower than usual.
4 True.
5 2008.
6 5.2.
7 b) Movement of rocks on a fault line.
8 London and Northern Ireland.
9 a) 0, b) 1.
10 £10,000 damage to church roof when a stone cross fell.

Knowledge test III (page 57)

1 Focus.
2 Epicentre.
3 Liquefaction.
4 Eurasian, Pacific and Philippine plates.
5 a) 5500 died, b) 250,000 homeless.
6 Earthquake drill.
7 Indo-Australian, Sunda and/or Eurasian plates.
8 9.2.
9 A tsunami up to 30 m high.
10 1.1 million.

Unit 2: Living in Our World

Theme A: People and Where They Live

Knowledge test I (page 78)

1 Grow rapidly.
2 Decrease.
3 A person who moves out of a country permanently.
4 Many migrants send money back home; fewer people to provide for.
5 Brain drain: fewer skilled workers.
6 a) 0–14, b) 65+.
7 a) Youth dependency.
8 Youth dependency.
9 Aged dependency.
10 Wise advice from elderly; grandparents can care for and about grandchildren.

Knowledge test II (page 78)

1 Site.
2 A hill top or inside a meander bend of a river.
3 Converging routes make it ideal for a market function so it is likely to grow.
4 An arrangement of settlements in order of size and importance.
5 True.
6 Range.
7 Port; seaside resort.
8 Central business district.
9 Inner city.
10 Cheaper land, close to motorway transport.

Knowledge test III (page 79)

1 Natural population increase.
2 Push factor.
3 Pull factor.
4 Bustees.
5 Four million.
6 Badly built, bustees, basic, beside roads, bulldozed, below poverty line.
7 Largely disused and derelict after declining Harland & Wolff shipyard no longer needed the majority of its original site.
8 Citigroup and Microsoft.
9 7500 apartments and townhouses.
10 1.5 km of waterside frontage; parks on former slipways of shipyard.

Theme B: Contrasts in World Development

Knowledge test I (page 93)

1 a) Social, b) social, c) economic.
2 Social.
3 MEDC.
4 MEDC.
5 HDI (human development index).
6 a) Life expectancy, b) adult literacy and school enrolment, c) GDP (gross domestic product).
7 LEDCs.
8 True.
9 True.
10 Two of disaster relief; water and education development projects; child sponsorship; research and campaigning.

Knowledge test II (page 94)

1 The way people, money, goods and ideas move around the world faster and more cheaply than ever before.
2 Multinational corporations.
3 Jobs and better standard of living for educated people; increased life expectancy and literacy.
4 Unemployment for unskilled people; shocking Western-style behaviour; political unrest.
5 Newly industrialising country.
6 False – NICs are more developed than LEDCs but less than MEDCs.
7 Trade.

8 MEDCs.
9 LEDCs.
10 LEDCs.

Knowledge test III (page 94)

1 Technology suited to the level of development in the area where it is used.
2 Jobs building the new boats; increased income from fishing.
3 Fewer trees cut down.
4 Better healthcare and education as the fishermen earned more money.
5 Petrol engines cause pollution; fibre glass is not produced locally.
6 Improved healthcare, education, transport and standard of living.
7 Chocolate, sugar, bananas, coffee, tea.
8 Voluntary; short-term.
9 To help a country become more developed.
10 One country gives aid directly to another country.

Theme C: Managing our Resources

Knowledge test I (page 107)

1 Farmland, water, fuel, housing.
2 Finite, or non-renewable.
3 Solar power, wind power.
4 Using up resources.
5 Carbon footprint.
6 Loft insulation, water heated by solar panels, cycling, using public transport.
7 Population growth leading to traffic congestion in narrow streets.
8 Improved public transport, park and ride, cycle paths, no free parking.
9 4000 fewer cars each day than in 1970, 70% of local trips made by tram, public transport use more than doubled since 1980, reducing congestion, fuel use and pollution.
10 Cars are still used in spite of all these measures.

Knowledge test II (page 107)

1 One million per month.
2 9.5% each year.
3 Air pollution in two-thirds of cities studied, four million fridges thrown away each year releasing CFCs, 10 million hectares of forest lost each year, tigers dying out.
4 750,000 deaths per year due to pollution, traffic congestion, water pollution.
5 Renewable methods.
6 Plants are processed or burnt to produce energy.
7 19%.
8 Wind farms in the sea damage plants and animals, some people dislike the look of turbines, some birds lose feeding grounds, difficult to store electricity, leads to drop in tourism.
9 No greenhouse gases, wind never runs out, electric cars could store electricity, need to import less fuel.
10 True.

Knowledge test III (page 107)

1 One of shortage of landfill sites, health and environmental concerns, EU laws.
2 c) Reusing
3 121,000 tonnes.
4 Incinerator, burning waste to produce energy.
5 Composting.
6 Two of increased leisure time, increased disposable income, cheaper travel, increased health/wealth of pensioners.
7 Hotels profits sent to owners overseas, migrant workers send money home.
8 Ecotourism.
9 Three of clouded leopards, tigers, gibbons, elephants.
10 Trekking on foot and boating.

Answers to Get Active

Page	Task	Answer
2	1	(A) percolation, (B) interception by vegetation, (C) precipitation, (D) through-flow, (E) infiltration, (F) evaporation, (G) groundwater flow, (H) river discharge
	2	(A) source, (B) mouth, (C) watershed, (D) tributary, (E) drainage basin, (F) confluence
4		Near source: steep, narrow, shallow, low, large angular Near mouth: gentle, wide, deep, high, small rounded
6 top		(a) erosion, (b) transportation, suspension, (c) erosion, (d) deposition, (e) transportation, traction or saltation, (f) erosion, attrition
6 bottom		(1) undercuts, (2) overhanging hard rock falls into plunge pool, (3) hydraulic action and abrasion make a plunge pool, (4) waterfall moves backwards, (5) soft rock, (6) hard rock
8	1	Inside bend: slower flow, less energy, deposition, slip-off slope, shallow water. Outside bend: faster flow, lots of energy, erosion, river cliff, deep water
	2	Mirror image of Figure 10, page 7
	4	White
9	1	(A) bluff, (B) floodplain, (C) river channel, (D) deposition
	2	(a) pale green, (b) less than 10 m high, (c) no contours across the floodplain, (d) village settlement, recreation, land suitable for grazing, (e) playing field flooded, town and caravan park flooded meaning people may be evacuated, main road along coast flooded and damaged, houses along the roads at the edge of the floodplain flooded, farmers move animals from floodplain to higher ground
11		(A) higher wave, (B) waves close together, (C) strong backwash pulls sediment down the beach, (D) waves far apart, (E) lower wave, (F) strong swash pushes sediment up beach, (G) water sinks into beach reducing backwash
12	1	(a) abrasion, (b) attrition and (c) hydraulic action
	2	1b, 2d, 3a, 4c

Page	Task	Answer
14	1	1d, 2a, 3c, 4b
	2	1c, 2a, 3d, 4b
15 top		1d, 2b, 3c, 4a
15 bottom	1	a3, b4, c2, d1
	2	Towards S/SW
17		–, –, –, +, –, –, +, –, +, –
18		a3S, b5H, c2H, d6S, e7H, f4S, g1H
20		(a) residential and tourism, (b) tourism and industry, (c) transport and erosion, (d) industry, residential and erosion, (e) transport industry and deposition
22		(a) sea wall, (b) beach nourishment, (c) gabions, (d) groynes
26		Weather: (a), (b) and (e). Climate: (c), (d) and (f)
27		Temperature: max and min thermometer, °C. Precipitation: rain gauge, mm. Wind speed: anemometer, knots. Wind direction: wind vane, eight points of the compass
28		Thermometer: E, anemometer: D, rain gauge: A, wind vane: D
30		(1) Polar martime: cool, cold, wet. (2) Polar continental: hot, cold, dry. (3) Tropical continental: hot, mild, dry. (4) Tropical maritime: warm, mild, wet
32		(1) D–, (2) SA–, (3) D+, (4) WA–, (5) D–, (6) WA–, (7) D+, (8) SA+, (9) SA+
34	1	Retailers need to plan months ahead to stock up with disposable barbecues and garden chairs or warm coats and gloves. Road service need to order suitable quantities of salt for gritting roads in winter
37		Greenhouse effect: (a), (c1), (g). Global warming: (b), (c2), (d), (e), (f)
38		(a) –economy, (b) –society, (c) –economy/environment, (d) –environment, (e) –society/economy, (f) –economy, (g) +society/economy, (h) –society/economy, (i) +economy, (j) –society
44	2	(a) extrusive igneous, (b) basalt, (c) intrusive igneous, (d) granite, (e) sedimentary, (f) sandstone/limestone, (g) metamorphic, (h) marble
48	1	(1) oceanic plate, (2) ocean trench, (3) earthquake focus, (4) volcano, (5) continental plate, (6) mantle
	2	Both arrow heads should point towards centre of the diagram

Page	Task	Answer
52		(1) water rises to the sand surface and (2) the building topples over
59		True: (2), (3), (5), (7), (8). False: (1), (4), (6), (9)
60		The country people leave: –, + , +. The country people go to: –, +, +, +, –. The migrants: + , –, –, +, +
62		Services: –, –, –, +, –. Economy: +, –, + , –, +, –
63		1e, 2b, 3a, 4d, 5h, 6c, 7f, 8g
64 top		UK: 1B, 2C, 3A
64 bottom		Afghanistan: 1C, 2A, 3B
65		MEDC social costs: 1, 12; social benefits: 2, 10. MEDC economic costs: 3, 8, 9, 11. LEDC social costs: 4, 5, 13. LEDC economic costs: 6; economic benefits: 7
68		1b, 2d, 3e, 4a, 5c
72		Two caravan sites, golf course and route of Ulster Way
73		Push: 1, 3. Pull: 2, 4, 5
74		Big: four million people in Kolkata, growing at 32% in 10 years. Badly built: scrap wood or corrugated iron. Basic: few facilities, shared water taps, few toilets. Beside roads: because people can get buses to jobs. Bulldozed: if not on cheap land that no-one else wants. Below poverty line: average earnings £7–£24 per month
75		Housing: (a), (f). Employment: (b), (d), (e), (h). Environment: (c), (g), (i)
81		Social: 2, 3, 4, 6. Economic: 1, 5
83		Historical: 7, 9. Environmental 1, 6, 8. Dependence on primary activities: 5, 10. Debt: 2, 3. Politics: 4
84		(C) the organisation, (A) core aims, (B) action taken
86		Help: 2, 3, 5, 7, 8; Hinder: 1, 4, 6, 9. *Note: these answers depend on your point of view of what development should be about – some people might think enormous shopping centres and lots of mass consumption are not good things! For the purposes of your GCSE it is easiest to assume they are.*

Page	Task	Answer
87		(A) LEDC, (B) MEDC
88		1a, 2a, 3a, 4a, 5b, 6b
89	1	Positive (a), (b), (c), (e), (f). Negative (d), (g), (h)
	2	(a) economic, (b) economic, (c) environmental, (e) economic, (f) social
92		(1) voluntary, short term; (2) bilateral; (3) multilateral, long term, tied; (4) voluntary, long term
93		Positive: fewer people die …, medical care …, … education and jobs,… provide for families …, … lives made easier, saves lives in emergencies. Negative: aid may be unsuitable,… loans and debt …, doesn't reach those in most need,… corrupt governments …, local producers lose out …, people dependent on aid.
96		Caroline: (a), (d), (g), (h), (j). Gregory: (b), (c), (e), (f), (i)
100		Benefits: (b), (c), (g), (i). Problems: (a), (d), (e), (f), (h).
101	1	(a) landfill, (b) incineration/waste to energy, (c) recycling, (d) reusing, (e) reducing, (f) composting
	2	Reducing, reusing, recycling, composting, incineration/waste to energy, landfill
102		(a) reduce and reuse, (b) reduce and reuse, (c) recycle, (d) reduce, (e) recycle, (f) reduce, (g) recycle, (h) reduce and reuse
105		(1) –environmental, (2) –environmental, (3) +economic, (4) –cult, (5) –economic
106		(1) a, c, e, h; (2) b, f, g; (3) j, k, l; (4) d, i, j

Glossary

Abrasion/corrasion (1) The grinding of rock fragments carried by a river against the bed and banks of the river. (2) A process of erosion which occurs when a wave hits the coast and throws pebbles against the cliff face. These knock off small parts of the cliff causing undercutting

Aged dependency The proportion of people aged 65 or over in a population

Air mass A body of air with similar characteristics, e.g. temperature, humidity and air pressure

Anemometer An instrument which is used to measure wind speed

Anticyclone A weather system with high pressure at its centre

Appropriate technology Technology which is suited to the level of development in the area where it is used

Atmospheric pressure The weight of a column of air measured in millibars

Attrition A process of erosion where transported particles hit against each other making the particles smaller and more rounded

Barometer An instrument used to measure air pressure

Bilateral aid Resources are given directly from a rich 'donor' country to a poorer 'recipient' country. If conditions are placed on the way in which the money is used, then this becomes tied aid

Biodiversity Biodiversity is the variation of life forms within an ecosystem

Birth rate The number of live births per 1000 of a population per year

Carbon footprint A measure of the amount of carbon dioxide produced by a person, organisation or country in a given time

Central business district (CBD) The part of a town (or larger settlement) which is dominated by shops and offices and is usually close to its centre

Climate The average weather conditions of an area over a long period of time, e.g. 35 years

Cloud cover The amount of sky covered by cloud, measured in oktas

Cold front The zone where cold air comes behind warm air. The cold air undercuts the warm air forcing it to rise cool and condense

Confluence The point where two rivers meet

Constructive wave A wave with a strong swash and weak backwash which contributes deposition to a beach

Convection current Repetitive movements set up in the mantle due to heating by the core. These currents make the crust move

Core The centre of the Earth, found below the mantle. It is extremely hot and may be made of metal

Counterurbanisation The movement of people away from towns and cities to smaller towns, villages or areas in the countryside

Crust The upper layer of the Earth on which we live. It is solid but is split into sections called plates

Death rate The number of deaths per 1000 of a population per year

Deposition The dropping of material on the Earth's surface

Depression A weather system with low pressure at its centre

Destructive wave A wave with a strong backwash and weak swash which erodes a coast

Development The level of economic growth and wealth of a country. The use of resources, natural and human, to achieve higher standards of living. This includes economic factors, social measures and issues such as freedom

Development gap The division between wealthy and poor areas, in particular the disparity between LEDCs and MEDCs

Discharge The amount of water in a river which is passing a certain point in a certain time. It is measured in cumecs (cubic metres per second)

Drainage basin An area of land drained by a river and all of its tributaries

Earthquake A tremor starting in the crust which causes shaking to be felt on the Earth's surface

Economic indicators Figures relating to the wealth and economy of a country

Ecotourism Otherwise known as green tourism. A sustainable form of tourism which involves protecting the environment and local way of life at the destination

Emigration The movement of people away from one country to another

Energy-from-waste plant Sometimes called incinerators, these use the parts of waste that burn to generate electricity

Epicentre The first place on the Earth's surface to feel shockwaves from an earthquake. It is directly above the focus

Erosion Wearing away of the landscape by the action of ice, water and wind

Fair trade A type of trade where producers in a poor country get a fair living wage for their product and which promotes environmental protection

Fault line A weak line in Earth's surface, where crust is moving, causing earthquake activity

Flooding A temporary covering by water of land which is normally dry

Focus The point of origin of an earthquake under the Earth's surface

Fossil fuel Any resource found in the Earth's crust which contains carbon and can be burnt to release heat, e.g. coal, oil or gas

Front The zone where two types of air mass meet

Function of a settlement The main reason why a settlement is there

Global warming The warming of the atmosphere, i.e. the increase over time in average annual global temperature. This is probably related to human activity through the release of greenhouse gases

Globalisation The way in which countries from all over the world are becoming linked by trade, ideas and technology

Greenhouse effect A natural process where our atmosphere traps heat. Some of the insolation that is absorbed by the Earth's surface is re-radiated to the atmosphere where it is held by the greenhouse gases – carbon dioxide, methane, nitrogen dioxide, CFCs and water vapour

Groundwater flow Water which is moving through the bedrock

Hard engineering A strategy to control a natural hazard which does not blend into the environment

High order function A good or service which is used infrequently, maybe because it is expensive

Human development index (HDI) A measure of development which combines measures of wealth, health and education, thus mixing social and economic indicators

Hydraulic action (1) A form of erosion caused by the force of moving water. It undercuts riverbanks on the outside of meanders and forces air into cracks in exposed rocks in waterfalls. (2) The process whereby soft rocks are washed away by the sea. Air trapped in cracks by the force of water can widen cracks causing sections of cliff to break away from the cliff face

Immigration The inward movement of people to a country from another

Infiltration The movement of water into the soil

Interception The process whereby precipitation is prevented from falling onto the ground by plants. It slows run-off and reduces the risk of flash flooding

Landfill site A large hole in the ground into which rubbish is dumped

Lava plateau This is a flat, wide surface that is formed when lava comes out of the ground and spreads out quickly

LEDC A less economically developed country, often recognised by its poverty and a low standard of living

Liquefaction The process of solid soil turning to liquid mud caused by shaking during an earthquake bringing water to the surface

Long-shore drift The process whereby beach material moves along a coastline, caused by waves hitting the coast at an angle

Low order function Goods or services which are used regularly

Mantle The layer above the Earth's core. It makes up 80% of the Earth's mass. It behaves like liquid rock

Maximum thermometer An instrument used to measure the hottest temperature reached in a place

MEDC A more economically developed country, often recognised by its wealth and a high standard of living

Mid-ocean ridge Where two plates made of oceanic crust move apart, the magma of the mantle rises to fill the gap, causing the crust to rise and form a ridge

Migration The permanent or semi-permanent movement of people from one place of residence to another. Migration can be classified, for example into *forced*, e.g. due to war or famine, or *voluntary*, e.g. looking for better work

Millibar the unit used to measure air pressure

Minimum thermometer An instrument used to measure the coldest temperature reached in a place

Mouth The end of a river where it meets the sea, ocean or lake

Multilateral aid MEDCs give money to an international organisation, like the World Bank, which then redistributes the money to LEDCs

Natural increase The positive difference between the birth rate and the death rate. For example, birth rate = 41 and death rate = 20, then natural increase = 21 per thousand

Newly industrialised country (NIC) A country which experienced rapid economic growth since the 1980s, e.g. South Korea

Ocean trench A feature of a destructive plate margin which involves oceanic crust. Where the oceanic crust is forced down into the mantle it sinks below its normal level to create a deep trench in the ocean

Percolation The movement of water from the soil into the bedrock

Plate margin/boundary A zone where two plates meet. Plate boundaries may be described as constructive, destructive, conservative or collision

Plates Sections of the Earth's crust which are constantly moving due to convection currents set up in the mantle

Population structure The way in which a population is made up of males and females of different ages

Precipitation A form of moisture in the atmosphere, such as rainfall, sleet, snow and fog

Primary activity An activity which uses the Earth's resources as a way of making money, e.g. farming fishing, agriculture, mining

Pull factor Any attractive/positive aspect or quality of a place which attracts (pulls) migrants to it

Push factors Any negative aspect or quality of a place which causes people to leave it

Quality of life A measure of a person's emotional, social and physical well-being

Quaternary activity An activity which focuses on the research and development of products

Rain gauge An instrument which catches and measures precipitation

Range The maximum distance which people are prepared to travel to get to a good/service

Renewable energy A sustainable source of electricity production such as wind, solar or biofuels

Resource Anything that we use and rely on. A renewable resource is a sustainable resource which can be used over and over again without running out. Resource depletion is when a resource is used at a faster rate than it is being replaced, causing it to run out. A non-renewable resource is a resource which cannot be replaced in geological time once it has been used

Richter scale A scale between 0 and 9 which measures the strength of an earthquake

Rock types Geologists divide rocks into three rock types, depending on how the rock was made: igneous, sedimentary and metamorphic

Rural–urban fringe An area on the outskirts of a city beyond the suburbs where there is a mixture of rural and urban land uses

Saltation The bouncing of medium-sized load along a river bed or the seabed

Satellite image A photograph taken from space

Secondary activity An activity which involves the manufacturing or building of a product which is sold to make money. The processing of a raw material

Seismograph An instrument designed to measure the energy released by earthquakes

Settlement hierarchy The arrangement of settlements in order of importance from small to large, eg ranging from an individual farmstead to a megalopolis

Settlement A place where people live

Shanty town A characteristic of LEDC cities; an area within them of unplanned poor quality housing which lacks basic services like clean water

Site The physical characteristics of the land on which a settlement is located, e.g. wet point, bridging point or defensive site

Social indicators Figures relating to the quality of life within a country

Soft engineering A strategy to control a natural hazard which does blend into the environment so is often sustainable

Solution/corrosion The process by which water (in river or sea) reacts chemically with soluble minerals in the rocks and dissolves them

Source The starting point of a river, it may be a lake, glacier or marsh

Sphere of influence The area served by a settlement

Subduction zone An area where crust is being forced down into the mantle

Surface run-off/overland flow Water which is moving over the surface of the land

Suspension The transportation of the smallest load, e.g. fine sand and clay which is held up continually within river or seawater

Sustainable A way of using resources so that they are not destroyed but remain available for others to use in the future

Synoptic chart A weather map which shows the weather as symbols over an area

Temperature The hotness or coldness of the air in relation to weather. It is usually measured in degrees Celsius

Tertiary activity An activity where a service is provided

Threshold of a product or service The minimum number of people needed to support a good/service

Through-flow Water which is moving through the soil

Tied aid When conditions are placed by the 'donor' on the way that money or resources are used by the 'recipient' country

Traction The rolling of large rocks along a river or seabed

Trade The business of buying, selling and distributing goods and services

Transportation The movement of material across the Earth's surface

Tributary A stream which flows into a larger river

Tsunami Large waves caused by underwater earthquakes

UNESCO The United Nations Educational, Scientific and Cultural Organisation (UNESCO) is part of the United Nations. It promotes peace and security through international collaboration in education, science, and culture

Urban sprawl The unplanned growth of a city into the nearby countryside

Urbanisation An increase in the proportion of a country's population who live in urban areas

Volcanic plug A landform made from the hardened vent material from inside a volcano. This material is exposed when the surrounding volcano is eroded away

Volcano A cone-shaped mountain built up from hardened ash and lava, from which molten material erupts onto the Earth's surface

Voluntary aid The general public in MEDCs give money to voluntary organisations like Oxfam, which use the money to fund development projects

Warm front The zone where warm air comes behind cold air

Waste hierarchy The arrangement of waste disposal options in order of sustainability

Water cycle The continuous circulation of water between land, sea and air

Watershed The boundary between drainage basins, it is often a ridge of high land

Weather The day-to-day condition of the atmosphere. The main elements of weather include rainfall, temperature, wind speed and direction, cloud type and cover and air pressure

Wetlands Marshes, bogs and lakes which provide rich habitats

Wind The movement of air within the atmosphere

Wind direction The geographical direction (compass point) from which a wind blows

Wind speed The speed at which air is flowing. It can be measured in knots

World Health Organisation The World Health Organisation (WHO) is an agency of the UN (United Nations) that coordinates public health around the world, with its headquarters in Geneva, Switzerland

Youth dependency The proportion of people aged 15 or under in a population